Radio Tubes and Boxes
of the 1920's

SONORAN
PUBLISHING, LLC
Chandler, Arizona

Sonoran Publishing, LLC, Chandler, Arizona 85226
©1999 by George A. Fathauer. All rights reserved.
First printing July 1999

Printed in the United States of America

Library of Congress Cataloging-in-Publication Data

Fathauer, George A., 1954-
    Radio tubes and boxes of the 1920's / George A. Fathauer.
       p.   cm.
    Includes bibliographical references (p.).
    ISBN 1-886606-13-7 (pbk.)
     1. Vacuum-tubes--Collectors and collecting--United States.
  2. Radio supplies industry--United States--History.
  TK6565.V3F37 1999
  621.384'132'075--dc21   99-32246
                                             CIP

CONTENTS

DeForest Double-Wing Audion, c. 1915
DeForest "spherical" audions were the first commercially produced triode vacuum tubes in the United States. (Joe Knight Collection)

ACKNOWLEDGMENTS

The idea of producing a picture book of vacuum tubes for collectors has been mentioned to me by various individuals from time to time, and being in the book publishing business I would always offer to publish the book if someone would author it. No one stepped forward, however, and one day as I was viewing a nice collection of 201/A tubes in unusual brands I realized that it provided an opportunity to get my camera and start shooting photos, with the thought that maybe I could eventually accumulate enough suitable material to do the book myself.

I knew that the book would never become a reality without contributions from other collectors, which proved to be no obstacle as the radio collecting community quickly rallied around me with offers of tubes, photos, magazine ads and other literature. I am especially indebted to Joe Knight and Larry Daniel, who photographed many tubes from their collections time and again for me, and provided literature and magazine ads as well. I am also indebted to local collectors Martin Bergan, George H. Fathauer, Les Rayner, Clyde Watson, Debra Krol, Dennis Craft and Bill Lettow for making their collections available to me to photograph. I would like to thank Bro. Patrick Dowd for his list of 537 vintage 01/A tube brands, Alan Douglas for providing company brochures of several of the early independent tube makers and Lauren Peckham for help with dating and other information on some of the tubes. Finally, I would like to recognize Steve York for amassing the collection of hundreds of 201/A tubes in various brands, which provided the original inspiration for this book.

Radiotron

REG. U.S. PAT. OFF.

RADIOTRON
MODEL UV-201-A
PATENTED

Licensed for Amateur or
Experimental Use Only

Radiola
IV

In high quality
receiving sets, the
vacuum tubes —
the heart of their
fine performance
— bear the name
Radiotron and
the **RCA** mark.
Be sure to look for
this identification
when you replace
your tubes.

This symbol
of quality
is your
protection

Radiola
Grand

Radiotron	WD-11
Radiotron	WD-12
Radiotron	UV-199
Radiotron	UV-200
Radiotron	UV-201-A

Radio Corporation of America

Sales Offices:

233 Broadway, New York   10 So. La Salle St., Chicago, Ill.   433 California St., San Francisco, Cal.

*Send for the free booklet that describes
all Radiotrons and Radiolas.*

RADIO CORPORATION OF AMERICA
Dept. 212 (Address office nearest you.)
Please send me your free Radio Booklet.

Name _____

Street Address _____

City _____   R.F.D. _____

State _____

Radiola

REG. U.S. PAT. OFF.

The American Magazine, 1923

INTRODUCTION

The 1920's was a decade that witnessed dramatic change in radio communications. Prior to 1920 radio in its infancy was used primarily for military, weather, ships at sea and by a small number of radio amateurs. On November 2, 1920 radio station KDKA, operating from a small studio atop one of the Westinghouse buildings in East Pittsburgh, began regular broadcasts with coverage of the Harding-Cox presidential election returns. The concept of using radio broadcasts for the purpose of informing and entertaining the general public was born, and the following year found The Radio Corporation of America marketing receiving sets to the general public for home use. Radio stations sprung up across America almost overnight and dozens of new radio manufacturers appeared with hopes of cashing in on the new trend.

Others tried to profit from the radio boom by selling parts and supplies for radios. Replacement tubes appeared under every brand name imaginable — hundreds of them. Many of these new companies were bootleggers, or in other words didn't have the necessary RCA license to use their patented technology in the production of vacuum tubes. These manufacturers often appeared and disappeared almost overnight, only to re-open in a new location under a different name.

Small independent tube manufacturers and sellers used colorful packaging and persuasive advertising to sell their tubes to the public. Slogans such as "Renowned for Sound" (Cornell), "A Triumph in Radio Reception" (Marathon) or "Nuther Tube if This Don't Work" (By Heck) were employed along with marketing gimmicks such as colored glass.

Eventually the good times came to an end for most of these companies. The stock market crash of 1929 signaled the beginning of the great depression, while mass production techniques employed by companies like Philco greatly lowered the price of radios, making it difficult for others to compete. Most of the radio manufacturers could not adapt, and only a few of them survived past 1933. RCA, meanwhile, relentlessly pursued through the courts anyone who infringed on their patents, hastening the end of the tube bootleggers. By 1932 only twenty companies existed that were licensed by RCA for vacuum tube manufacture, and some of them didn't survive long enough to produce tubes under the new RMA numbering system introduced in 1933.

This book attempts to capture the excitement and confusion of the early days of radio, through a pictorial presentation of the radio replacement tubes of the era. Examples of these tubes can still be found at antique radio swap meets, and have become very collectible. Many early vacuum tubes are also found which are special purpose types not used in home radios, such as Western Electric tubes, and are not covered in this book. Neither is it in the scope of this book to cover the early development of the vacuum tube prior to 1920, or non- U.S. made tubes. Other books have been published which address these areas, and can be found listed in the bibliography at the back of this book.

Dating Tubes

While it would be nice to be able to show a date in the caption of every tube photo in this book, the fact that common types like the 201A were in production for many years, even decades, together with the fact that little or nothing is known of the miriads of manufacturers and sellers that came and went during this period, render it an impossible task to arrive at precise dates of manufacture for every tube found. For this reason, dates are only shown for those tubes where information is available, and the date given denotes the year when production was begun on that particular type by that particular manufacturer.

There are, however, some general guidelines based on historical development of tubes that can be helpful in making an educated guess at a date. By knowing when the various tube types came into production, and observing some simple characteristics of construction, a ballpark date can usually be determined for a tube. The following illustrations show the dates of introduction of common U.S. tube types, as well as the evolution of envelope shapes and base types.

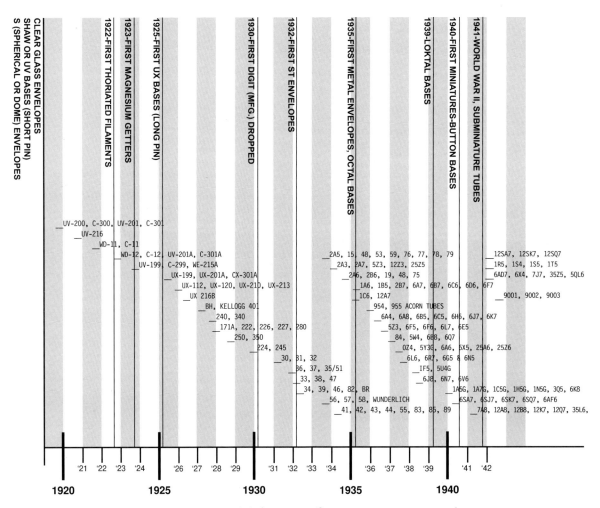

U.S. Vacuum Tube Development: 1921-1941

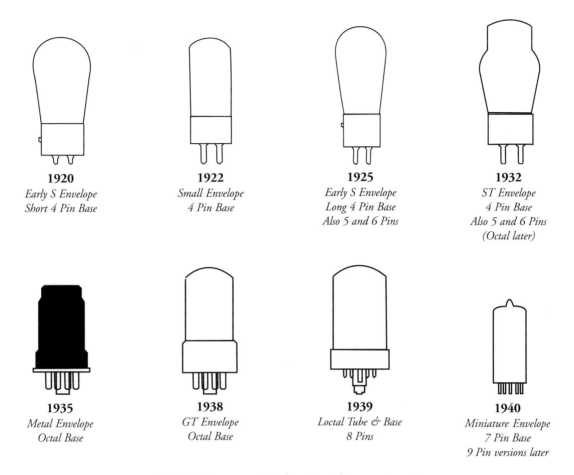

1920
Early S Envelope
Short 4 Pin Base

1922
Small Envelope
4 Pin Base

1925
Early S Envelope
Long 4 Pin Base
Also 5 and 6 Pins

1932
ST Envelope
4 Pin Base
Also 5 and 6 Pins
(Octal later)

1935
Metal Envelope
Octal Base

1938
GT Envelope
Octal Base

1939
Loctal Tube & Base
8 Pins

1940
Miniature Envelope
7 Pin Base
9 Pin versions later

U.S. Vacuum Tube Configurations

In addition, it is helpful to bear in mind that type 201/A, WD-11, and 199 or equivalent tubes usually had brass bases from 1920 to 1924. RCA introduced the molded bakelite base in 1924, and the other manufacturers soon followed suit. RCA also switched to "tipless" glass construction in 1924, although some of the other manufacturers, particularly Arcturus, were still making "tipped" tubes as late as 1929. The UV, or "short pin" type base can also indicate an early date, as RCA ceased production of these in August, 1925 in favor of the UX or "long pin" base, as collectors call it. Additional help with dating can be obtained by consulting the many excellent books listed in the bibliography found at the back of this book.

Identification

Prior to 1934 tube manufacturers used any numbers or letters they wished to designate their tubes, although most followed the RCA numbering scheme. In 1934 the Radio Manufacturers Association (RMA) began standardizing the process, introducing a system for receiving tubes where the initial number on a tube denoted the filament voltage, followed by a letter or letters that indicated the application, followed by another number which indicated the number of useable elements. A 2A3, for example, was a power triode with a 2 1/2 volt filament.

This book covers primarily tube types that were introduced during the 1920s, with a few exceptions, and in all cases types that preceded the RMA numbering system. Some of the tubes pictured herein may have been manufactured later than the 1920s, as their ST bulbs will attest, but still represent types that found their genesis in the 1920s, and are included here for the purpose of observing the continuing evolution of these types and their packaging. A few tubes are also pictured here that were developed prior to 1920, and have been included because they were still being sold into the early 20s and represent what was available in that decade.

The tube designations found in the captions of the photos are given exactly as shown on the tube or box, even to the degree of observing hyphens and spaces where they appear in the tube numbers. In other words, a type 01A/201A may be given as 01A, UV-201, UV-201A, UX-201A, C-301, CX-301A, GX-201A, 2976 (WLS), A (CeCo), etc. While these may all be electrically equivalent, the unique way each manufacturer or seller designated its tubes is part of the history and confusion of the era, and it is not the author's intent to try and standardize this.

Manufacturer's names and locations changed frequently as they came and went, were sold or merged with others. Therefore CeCo tubes, for example, may be identified in their captions as being made by the C.E. Manufacturing Co., The CeCo Manufacturing Co., Inc., or the Gold Seal Manufacturing Co., Inc., depending on which incarnation of the company was in business when that particular tube was made. Company locations, where given, reflect the information found on the tube carton, or the address shown in the various trade directories of the day. Oftentimes these addresses represented sales offices or corporate headquarters, not the location of manufacture.

In some cases, the author has tried to assist the reader by offering a helpful note in the caption. For example, the CeCo type A caption contains the comment that it is "equivalent to the UX-201A." Since this is primarily a picture book, however, it is not in the scope of this project to offer electrical data and a complete history for every tube shown herein. Again, there are many helpful references listed in the bibliography, and it would behoove the collector to begin building a personal library of these books.

RADIO TUBES AND BOXES

ACE A
(Kenneth and Debra Krol Collection)

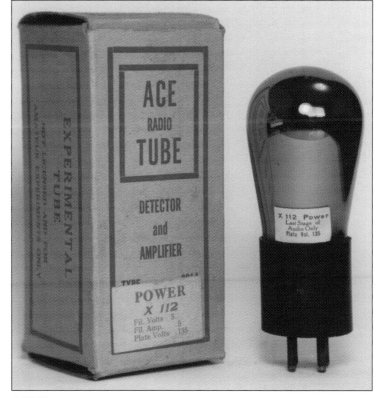

ACE X 112
These had gold color glass. (Joe Knight Collection)

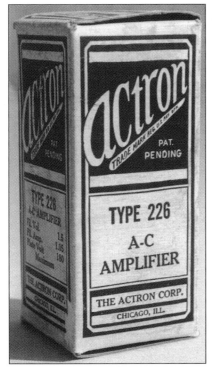

Actron 226
The Actron Corporation, Chicago, Illinois
(Larry Daniel Collection)

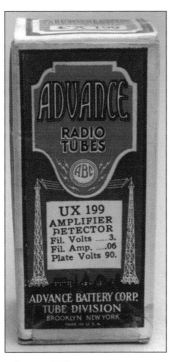

Advance UX 199
Advance Battery Corporation,
Tube Division, Brooklyn, N.Y.
(Clyde Watson Collection)

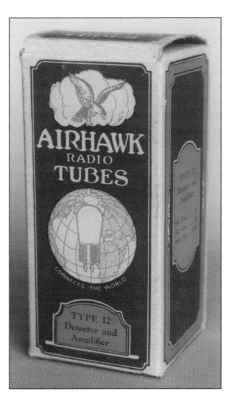

Airhawk 12

Air-King Y-227

Air Scout 227
(Clyde Watson Collection)

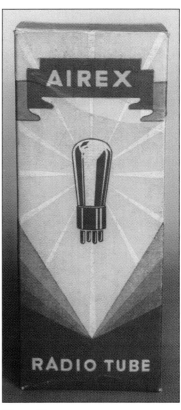

Airex 45
New York, Los Angeles
(Clyde Watson Collection)

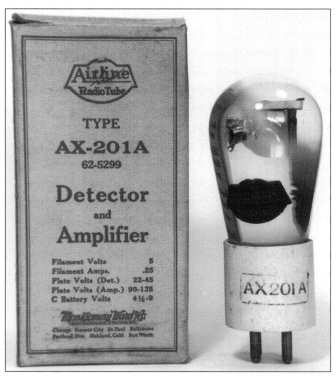

Airline AX-201A
Montgomery Ward & Co., Chicago. Note the ceramic base. (Joe Knight Collection)

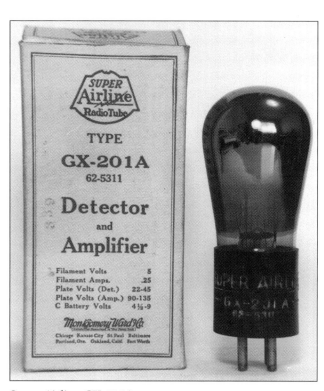

Super Airline GX-201A
Montgomery Ward & Co., Chicago. These had gold color glass. (Joe Knight Collection)

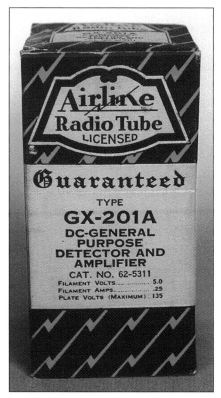

Airline GX-201A
Montgomery Ward & Co., Chicago

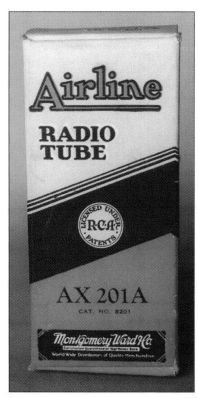

Airline AX 201A
Montgomery Ward & Co., Chicago

Super Airline 01AA
Montgomery Ward & Co., Chicago. The AA suffix signified a low current filament.

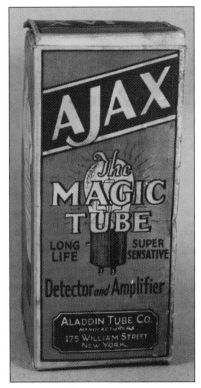

Ajax 227E
Aladdin Tube Co., New York, N.Y.
(Larry Daniel Collection)

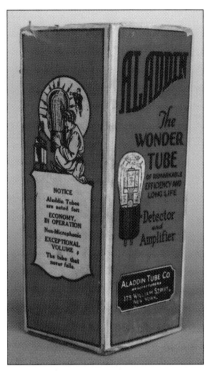

Aladdin 199X
Aladdin Tube Co., New York, N.Y.
(Kenneth and Debra Krol Collection)

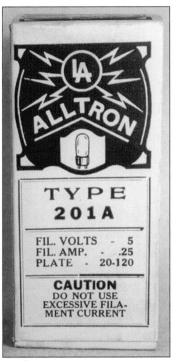

Alltron 201A
Alan Mfg. Co., Harrison, N.J.
(Larry Daniel Collection)

Alltron 201A
Alan Mfg. Co., Harrison, N.J.

American UX 201-A
Feldstern-McCusker Radio

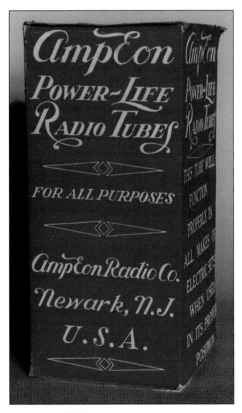

AmpEon UX-281
AmpEon Radio Co., Newark, N.J.

Amrad S-Tube, Type 4000, 1924
American Radio and Research Corporation, Medford Hillside, Ma.
(Joe Knight Collection)

Apex Audiotron A, 1924
Radio Tube Corporation, Newark, N.J. (Joe Knight Collection)

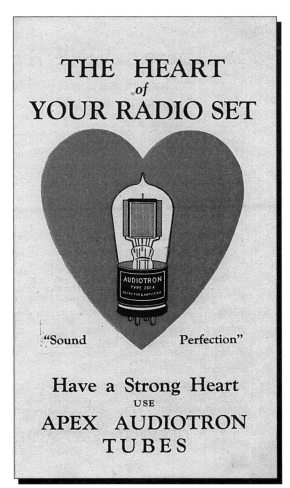

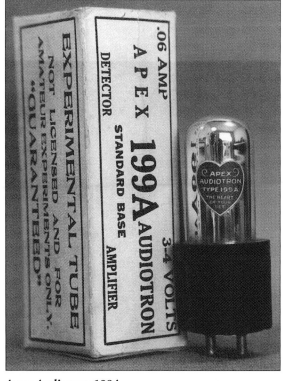

Apex Audiotron 199A
Radio Tube Corporation, Newark, N.J. (Joe Knight Collection)

Apex Brochure

APEX AUDIOTRONS
"Sound Perfection"

Apex Audiotrons have been on the market for a considerable length of time. They have been sold throughout the country and results have been highly gratifying. Apex Audiotrons are now being advertised nationally for the reason that the factory output has become large enough to take care of a great demand.

This is the first time that a manufacturer has sold a tube with a full guarantee. Apex Audiotrons may be relied upon at all times.

ALL TUBES ARE GUARANTEED TO WORK IN RADIO FREQUENCY. ESPECIALLY ADAPTED FOR NEUTRODYNE SETS.

The following tubes are now on sale:

PRICE

Type 201A—5 volts, .25 amperes . $4.00
    Amplifier and Detector

Type 199—3--4 volts, .06 amperes . $4.00
    Amplifier and Detector

Type 12—1½ volts, .25 amperes . . $4.00
    Platinum Filament—Amplifier and Detector

Type 200—5 volts, 1 ampere $4.00
    Detector Tube

LIST $4.00

APEX AUDIOTRON GUARANTEE

All Apex Audiotron tubes are guaranteed, and Dealers, as well as the manufacturers, will make replacement or refund the money on all tubes that prove unsatisfactory in any way. The only requirement is that the tube must not have been burned out.

"ATTENTION DEALERS"

The following Distributors supply "Apex Audiotron" tubes.

Radio Tube Exchange,
200 Broadway,
New York City

Radio Electric Co.,
218 Adams St.,
Scranton, Pa.

Radio Specialty Co.,
25 W. Broadway,
New York City

Standard Automotive Equipment Co.,
1074 Boylston St.,
Boston, Mass.

Radio Auto Supply Co.,
920 D St., N.W.,
Washington, D. C.

Baltimore Hub Wheel & Mfg. Co.,
Gay Street and Fallsway,
Baltimore, Md.

Wm. Spalding & Co., Inc.,
109-113 W. Jefferson St.,
Syracuse, N. Y.

Niles G. Plank,
17 So. Union St.,
Rochester, N. Y.

Radio Supply & Repair,
39 W. Adams St.,
Chicago, Ill.

Wellston Radio Co.,
1479 Hodiamont Ave.,
St. Louis, Mo.

Ridenour, Seaver & Kendig,
Caxton Bldg.,
Cleveland, Ohio

Globe Electric Co.,
Chamber of Commerce Bldg.,
Pittsburgh, Pa.,

W. P. Mussina,
1625 Barnard Ave.,
Waco, Texas

George H. Porell Co.,
453 Washington St.,
Boston, Mass.

The Radio Shop, Inc.,
26 So. Third St.,
Memphis, Tenn.

Eisenberg & Schaefer,
137 Market Street,
Philadelphia, Pa.

CANADIAN DISTRIBUTOR: Windsor Radio, Ltd., 26 Ferry Street, Windsor, Ontario

If your local dealer cannot supply you, order direct.

RADIO TUBE CORPORATION
70 HALSEY STREET   NEWARK, N. J.

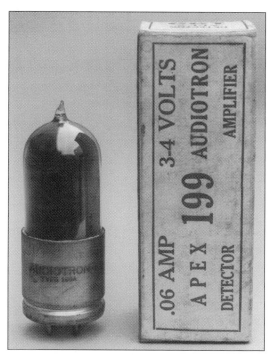

Apex Audiotron 199, 1924
Radio Tube Corporation, Newark, N.J.

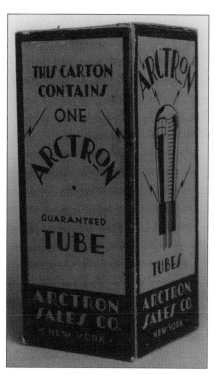

Arctron 224
Arctron Sales Co., New York

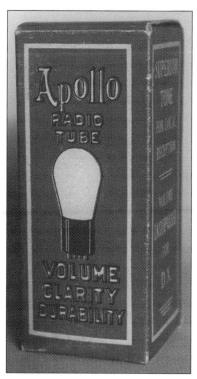

Apollo 227
Apollo Radio Co., Irvington, N.J.

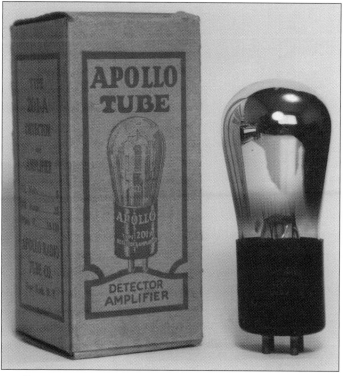

Apollo 201-A
Apollo Radio Tube Co., New York, N.Y. (Joe Knight Collection)

3 SIMPLE TUBE TESTS

That Are Boosting Business for Arcturus Dealers

There's no question about Arcturus' 7-second action when your customer holds the watch.

A 2-minute demonstration of Arcturus' clear, humless tone is more convincing than a 20-minute sales talk.

Show your customers that Arcturus Tubes hold the world's record for long life because they easily withstand 75% more current than they are designed for.

DEMONSTRATION has always been the most convincing way to sell a quality product.

You demonstrate sets and speakers...and sales follow.

Why not demonstrate tubes—one of the most important and profitable items you carry?

Last month we suggested this "Scientific Selection" idea to Arcturus Dealers. We illustrated three easy tube tests in our national advertising . . . showed the same tests in our trade paper advertising . . . distributed window and counter displays to the trade. Thousands of Arcturus Dealers adopted this test idea.

Now dealers everywhere report increasing Arcturus sales. And every Arcturus sale means better business, for Arcturus performance cuts service calls and keeps your customers satisfied.

Try these simple tests in your own store. Convince yourself and your customers of Arcturus quality. You can boost your business by pushing Arcturus *Blue* Radio Tubes . . . the tubes that sell on *proved performance*, not sales talk.

ARCTURUS RADIO TUBE COMPANY
Newark, N. J.
West Coast Representatives
UNIVERSAL AGENCIES
905 Mission Street, San Francisco, Calif.
201 Calo Bldg., Los Angeles, Calif.

ARCTURUS DETECTOR No. 127

ARCTURUS
RADIO   BLUE A-C LONG LIFE   TUBES

Arcturus 28, PZ and 127
Arcturus Radio Tube Co., Newark, N.J.
The PZ and 127 in this case were made for Crosley. (Joe Knight Collection)

Arcturus 124
Arcturus Radio Company, Newark, N.J. (Les Rayner
Collection)

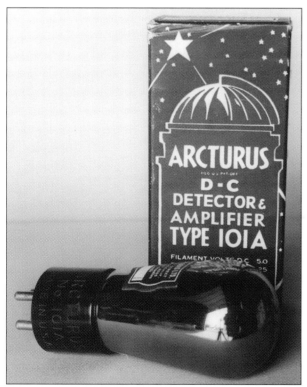

Arcturus 101A, 1929
Arcturus Radio Tube Co., Newark, N.J. Arcturus used blue
color glass.

Arcturus Wunderlich, 1932-33
Arcturus Radio Tube Co., Newark, N.J.
Full-wave detectors in three variations; 6.3 V. Automotive, and 2.5 V.
type A with 5 and 6 pins. (Martin Bergan Collection)

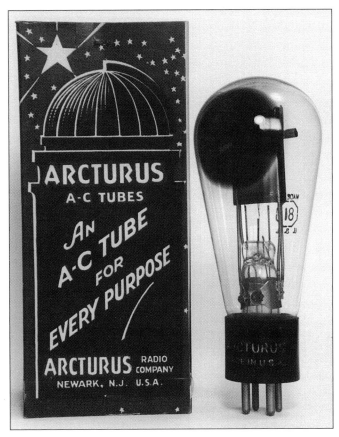

Arcturus 18
Arcturus Radio Company, Newark, N.J. (Joe Knight Collection)

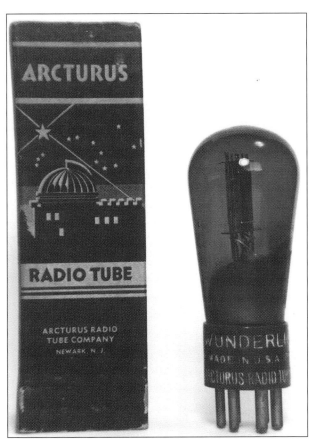

Arcturus Wunderlich, 1932
Arcturus Radio Tube Co., Newark, N.J. Full-wave dectector with 2.5 V. filament. (Joe Knight Collection)

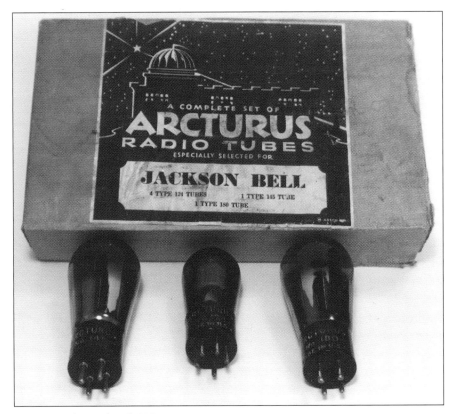

Arcturus 145, 124 and 180
Arcturus Radio Tube Co., Newark, N.J. Supplied in 6 tube repacement set for a Jackson Bell radio. (Joe Knight Collection)

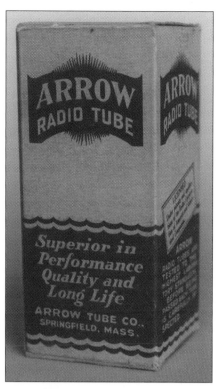

Aristocrat UX-201-A
(Clyde Watson Collection)

Armor CF512
Armstrong E&M Co., Inc., Newark, N.J.
(Larry Daniel Collection)

Arrow 280
Arrow Tube Co., Springfield, Mass.

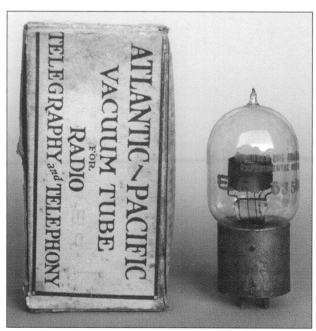

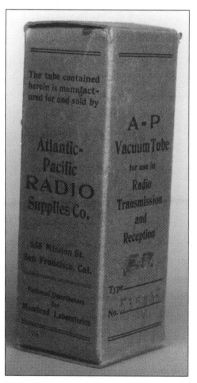

Atlantic-Pacific E.R., 1920
Atlantic-Pacific Radio Supplies Co., San Francisco,
California. "Electron Relay" detector.

A-P E.R., 1920
Atlantic-Pacific Radio Supplies Co.,
San Francisco, California. "Electron
Relay" detector.

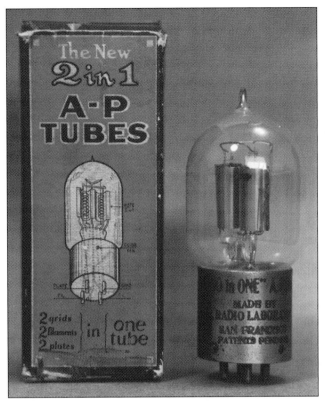

A-P 2 in 1, 1922
A-P Radio Laboratories, San Francisco, California.
This was a double-filament tube. (Joe Knight Collection)

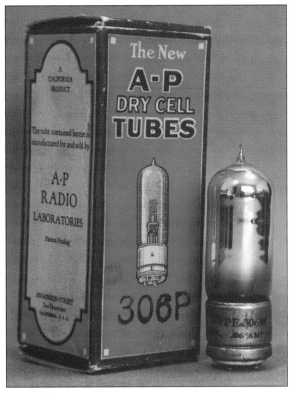

A-P 306P
A-P Radio Laboratories, San Francisco, California.
Equivalent to 199. (Joe Knight Collection)

Atwater Kent 607
Atwater Kent Manufacturing Co., Philadelphia, Pa. Gas rectifier tube for use
in A.K. model R and S battery eliminators. (Dennis Craft Collection)

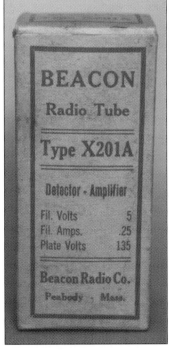

Beacon X201A
Beacon Radio Co., Peabody, Mass.

Beacon Blue B-201-A
Beacon Radio Co., Peabody, Ma. (Joe Knight Collection)

Beacon B-201-A
Beacon Radio Co., Peabody, Ma. (Clyde Watson Collection)

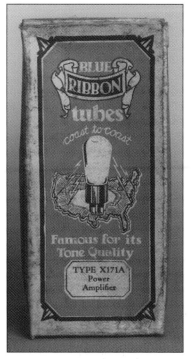

Blue Ribbon X171A

Radio News, **April 1925**

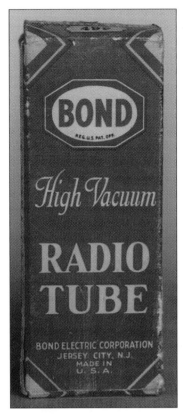

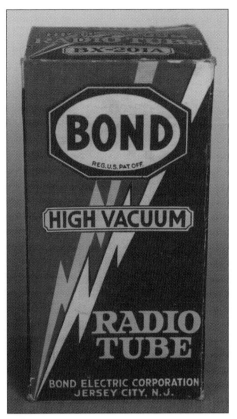

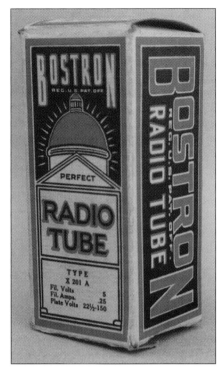

Bond 499
Bond Electric Corporation, Jersey City, N.J. (Clyde Watson Collection)

Bond BX-201A
Bond Electric Corporation, Jersey City, N.J.

Bostron X 201 A
(Larry Daniel Collection)

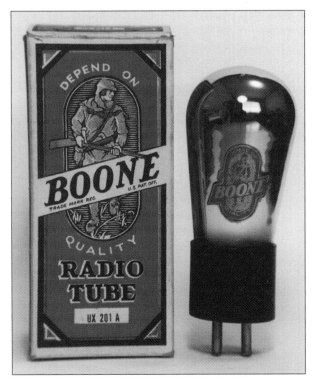

Boone UX 201 A
The Daniel Boone Co., New York, N.Y. (Joe Knight Collection)

Bright Star B.Y. 224
Bright Star Battery Co., Hoboken, N.J.

The Finest Radio Tube in the World!

NOT AN IMITATION OF ANY ORDINARY TUBE
Vastly Superior with Exclusive Features

True Blue Tubes have sterling silver contact points which prevent resistance losses from corroded tube prongs. A radio receiver delivers the full program with *True Blue's*.

Because of the special and exclusive construction, *True Blue* Radio Tubes *eliminate Microphonic noises*. Vibration does not distort radio programs when the radio receiver is equipped with *True Blue* Tubes. Neither are sponge rubber mountings necessary.

The genuine Bakelite base of *True Blue* Tubes is a mahogany color—a color that does not have any substance in it that might act as a conductor and cause weakened reception.

True Blue Radio Tubes reproduce all tone frequencies while operating with but four volts; yet for greater volume or longer distance the voltage may be varied to six volts.

The plate voltage of *True Blue* tubes as a Detector is 20-40, Amplifier 40-150. The filament consumption is only ¼ ampere.

As a result of all these features—and others—*True Blue* Radio Tubes make any radio set produce a clearer, sweeter tone than it would produce with any other radio tubes—and in greater volume than is possible with any other radio tubes.

The filament in a *True Blue* Radio Tube will last two to three times longer than any other radio tube on the market—making *True Blue* the most economical radio tube, regardless of the fact that its first cost—$6—is more than for ordinary tubes.

Manufactured by BRIGHTSON LABORATORIES, Inc.

GEORGE E. BRIGHTSON
President
Founder of Sonora
Phonograph Co.

True Blue

Northwest corner
Waldorf-Astoria Hotel
16 West 34th Street
New York, N. Y.

Brightson True Blue, 1925
Brightson Laboratories Inc., New York. Equivalent to 201A, in 5 tube set. (Joe Knight Collection)

Brightson True Blue, 1925
Brightson Laboratories Inc., New York. Equivalent to 201A. (Martin Bergan Collection)

Brightson True Blue, 1925
Brightson Laboratories Inc., New York, N.Y. Equivalent to 199. (Martin Bergan Collection)

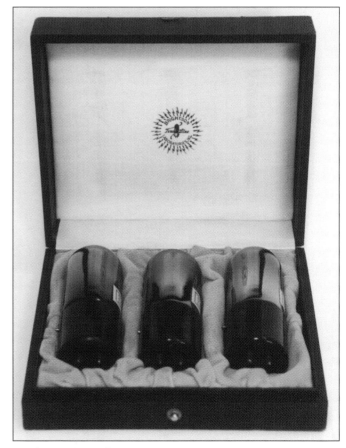

Brightson True Blue, 1925
Brightson Laboratories Inc., New York. Equivalent to 201A, in 3 tube set. (Joe Knight Collection)

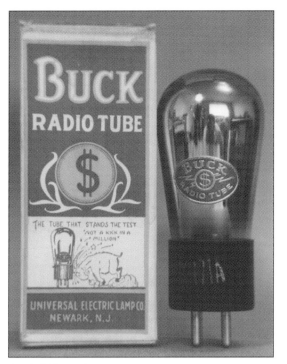

Buck UX 171A
Universal Electric Lamp Co., Newark, N.J. (Joe Knight Collection)

By Heck UX 201A
(Larry Daniel Collection)

By Heck X112A

Brook-Rad UX 112

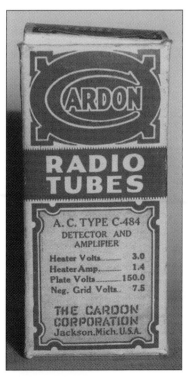

Cardon C-484, 1928-29
The Cardon Corporation, Jackson, Michigan. Non-standard, 3 V. filament.

Cardon C-586
The Cardon Corporation, Jackson Michigan. Equivalent to 50. (Kenneth and Debra Krol Collection)

C-R-A Sky-Sweeper X201A
Charles R. Ablett Co., New York, N.Y.
(Kenneth and Debra Krol Collection)

CeCo A, 1925
*C.E. Manufacturing Co., Inc. Providence, R.I. Equivalent
to UX-201A. (Joe Knight Collection)*

CeCo 99V
*Gold Seal Manufacturing
Co., Inc., East Newark, N.J.*

CeCo B/199
*C.E. Manufacturing Co., Inc. Providence, R.I. (Joe
Knight Collection)*

CeCo K, 1927
*C.E. Manufacturing Co., Inc. Providence, R.I. Type UX-201A
for RF use.*

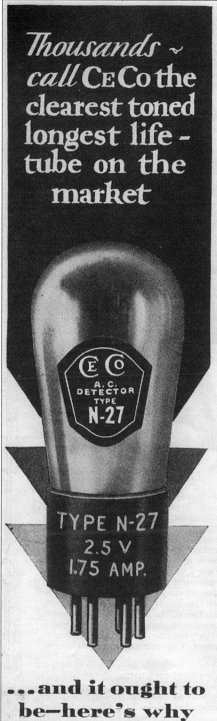

Thousands ~ call CeCo the clearest toned longest life - tube on the market

TYPE N-27
2.5 V
1.75 AMP.

...and it ought to be—here's why

THE finest materials skillfully employed by able craftsmen and the foremost laboratory experts give CeCo marked advantages over other tubes. Not only a purer, clearer, more pleasing tone, but a longer, more serviceable life.

CeCo Mfg Co., Inc. Providence, R. I.

CeCo AC Pentode, 1930
C.E. Manufacturing Co., Providence, R.I. RF Amplifier with connection to "space-charge" grid made through binding post on side of base. (Les Rayner Collection)

Champion X 201A
Champion Radio Works, Danvers, Ma. (Larry Daniel Collection)

Champion 224
Champion Radio Works, Danvers, Ma.

Radio, **January 1929**

Radio News, November 1930

Cleartron C.T. 199
Cleartron Vacuum Tube Co., New York, N.Y.

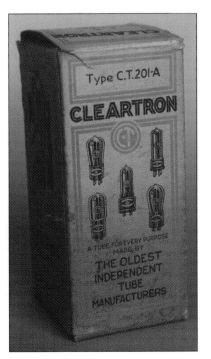

Cleartron C.T. 201-A
Cleartron Vacuum Tube Co., New York, N.Y.

Cleartron C.T.X. 201-A
Cleartron Vacuum Tube Co., New York, N.Y.

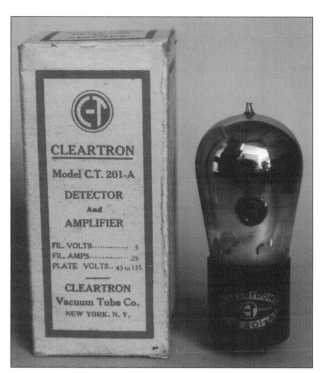

Cleartron C.T. 201-A
Cleartron Vacuum Tube Co., New York, N.Y.

Clover UY 227
Clover Battery Corp., Brooklyn,
N.Y. (Larry Daniel Collection)

Columbia 201A
(Larry Daniel Collection)

Commander 280

Concert Master 201A
Continental Corporation, Chicago

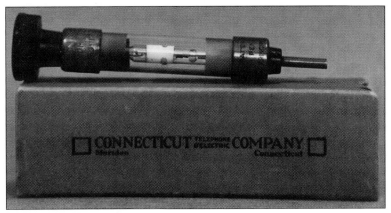

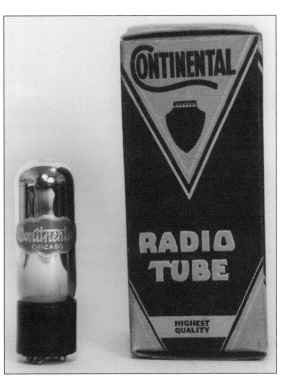

Connecticut J-117, 1921
Connecticut Telephone & Electric Co., Meriden, Ct. This detector tube was positioned inside a field coil and "tuned" by the bakelite knob on the end of the tube. (Joe Knight Collection)

Continental 199
Continental Corporation, Chicago, Illinois (Joe Knight Collection)

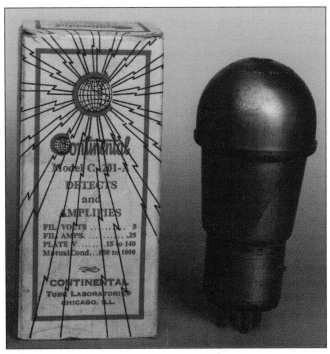

Continental C. 201-A
Continental Tube Laboratories, Chicago. Note copper shielding.

Cornell 227
Cornell Radio Laboratory, New York, N.Y.

Cornell C-201-A
Cornell Radio Laboratory, New York, N.Y. (Joe Knight Collection)

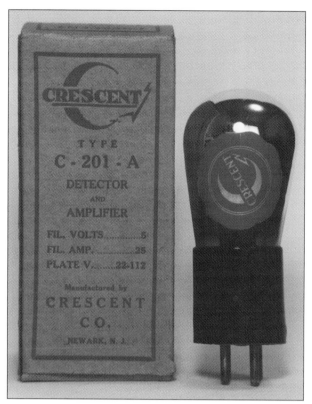

Crescent C-201-A
Crescent Co., Newark, N.J. (Joe Knight Collection)

Crescent 245
Crescent Radio Tubes (Kenneth and Debra Krol Collection)

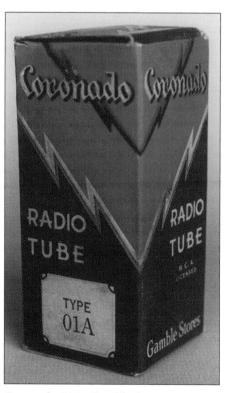

Coronado 01A, Gamble Stores

Crosley 01A
The Crosley Radio Corporation, Cincinnati, Ohio

Croydon 200A
*Gold Seal Electrical Co., New York,
N.Y. (Larry Daniel Collection)*

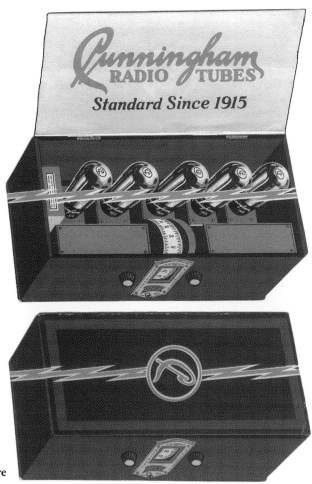

Cunningham Brochure

Cunningham C-300, 1920
*Audiotron Mfg. Co., San Francisco, California. The notation
"Trading as Audiotron Mfg. Co." was dropped in 1922 and
replaced with "E.T. Cunningham, Inc." (Joe Knight Collection)*

Cunningham C-301-A, 1923
*E.T. Cunningham, Inc., San Francisco, California (Joe
Knight Collection)*

Cunningham C-302, 1921
Audiotron Mfg. Co., San Francisco, California (Joe Knight Collection)

Cunningham C-300, 1924
E.T. Cunningham, Inc. San Francisco, California (Joe Knight Collection)

Cunningham CX-301-A, 1924
E.T. Cunningham, Inc., San Francisco, California (Joe Knight Collection)

Cunningham CX-371-A, 1926
E.T. Cunningham, Inc., San Francisco, California (Joe Knight Collection)

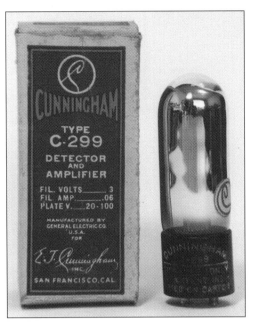

Cunningham C-299
E.T. Cunningham, Inc., San Francisco,
California (Joe Knight Collection)

Cunningham/RCA 01A
E.T. Cunningham Inc., Harrison, N.J.

Cunningham/RCA C-01A
E.T. Cunningham, Inc., Harrison, N.J.

Daven MU 20, 1925
Daven Radio Corporation, Newark, N.J. Shown with Daven resistance-coupled amplifier kit. (Larry Daniel Collection)

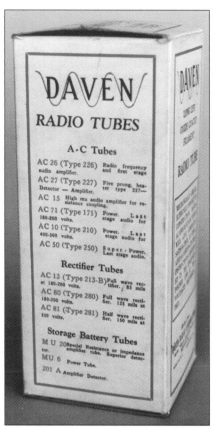

Daven AC 81
Daven Radio Corporation, Newark, N.J.
(Kenneth and Debra Krol Collection)

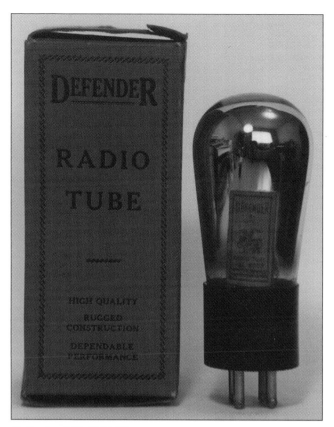

Defender 201 (#6505)
Sears, Roebuck and Co. (Joe Knight Collection)

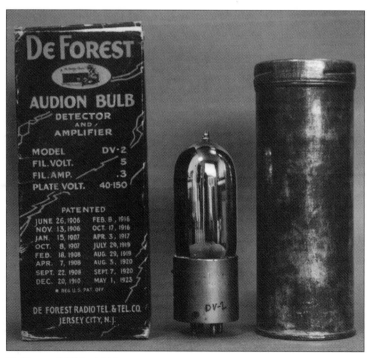

DeForest DV-2, 1923
DeForest Radio Tel. & Tel. Co., Jersey City, N.J. Tube originally came packaged in metal can, which was inside of box shown. (Joe Knight Collection)

DeForest DV-6, 1923
DeForest Radio Tel. & Tel. Co., Jersey City, N.J. (Les Rayner Collection)

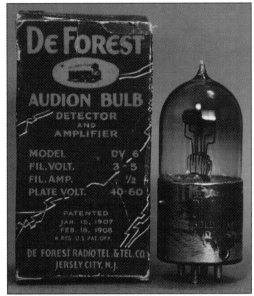

DeForest DV-6, 1923
DeForest Radio Tel. & Tel. Co., Jersey City, N.J. (Joe Knight Collection)

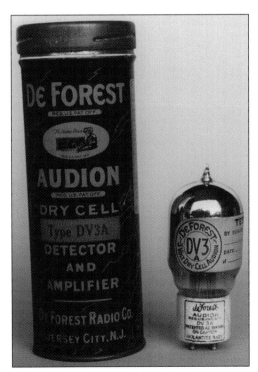

DeForest DV3A, 1925
DeForest Radio Co., Jersey City, N.J.

DeForest D-01A, 1925
DeForest Radio Co., Jersey City, N.J.

DeForest (left to right) DV5, DL3, DV3A, DV3, DV2 and DL2
DeForest Radio Co., Jersey City, N.J. The DV prefix indicated short pins, and the DL prefix long pins. (Joe Knight Collection)

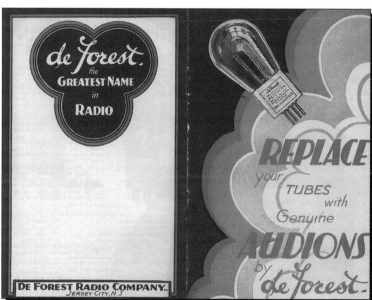

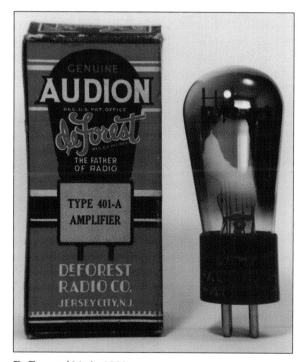

DeForest 401-A, 1930
DeForest Radio Co., Jersey City, N.J. (Joe Knight Collection)

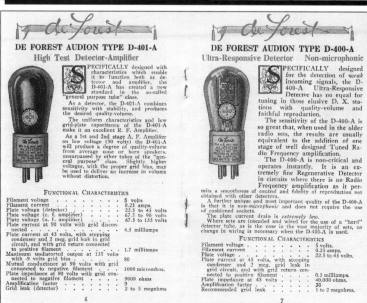

DE FOREST AUDION TYPE D-401-A
High Test Detector-Amplifier

SPECIFICALLY designed with characteristics which enable it to function both as detector and amplifier, the D-401-A has created a new standard in the so-called "general purpose tube" class.

As a detector, the D-401-A combines sensitivity with stability, and produces the desired quality-volume.

The uniform characteristics and low grid-plate capacitance of the D-401-A make it an excellent R. F. Amplifier.

As a 1st and 2nd stage A. F. Amplifier on low voltage (90 volts) the D-401-A will produce a degree of quality-volume from average cone or horn speakers, unsurpassed by other tubes of the "general purpose" class. Slightly higher voltages, with the proper grid bias, may be used to deliver an increase in volume without distortion.

FUNCTIONAL CHARACTERISTICS

Filament voltage	5 volts
Filament current	0.25 amps.
Plate voltage (detector)	22.5 to 45 volts
Plate voltage (r. f. amplifier) . .	67.5 to 90 volts
Plate voltage (a. f. amplifier) . .	67.5 to 135 volts
Plate current at 90 volts with grid disconnected	4.5 milliamps
Plate current at 45 volts, with stopping condenser and 2 meg. grid leak in grid circuit, and with grid return connected to positive filament	1.7 milliamps.
Maximum undistorted output at 135 volts with -9 volts grid bias	80
Mutual conductance at 90 volts with grid connected to negative filament . .	1000 micromhos.
Plate impedance at 90 volts with grid connected to negative filament	9000 ohms.
Amplification factor	9
Grid leak (detector)	2 to 5 megohms

6

DE FOREST AUDION TYPE D-400-A
Ultra-Responsive Detector   Non-microphonic

SPECIFICALLY designed for the detection of *weak* incoming signals, the D-400-A Ultra-Responsive Detector has no equal for tuning in those elusive D. X. stations with quality-volume and faithful reproduction.

The sensitivity of the D-400-A is so great that, when used in the older radio sets, the results are usually equivalent to the addition of one stage of well designed Tuned Radio Frequency amplification.

The D-400-A is non-critical and operates instantly. It is an extremely fine Regenerative Detector in circuits where there is no Radio Frequency amplification as it permits a smoothness of control and fidelity of reproduction not obtained with other detectors.

A further unique and most important quality of the D-400-A is that it is *non-microphonic* and does not require the use of cushioned sockets.

The plate current drain is *extremely low.*

Where sets are intended and wired for the use of a "hard" detector tube, as is the case in the vast majority of sets, no change in wiring is necessary when the D-400-A is used.

FUNCTIONAL CHARACTERISTICS

Filament voltage	5 volts.
Filament current	0.25 amps.
Plate voltage	22.5 to 45 volts.
Plate current at 45 volts, with stopping condenser and 2 meg. grid leak in grid circuit, and with grid return connected to positive filament	0.5 milliamps.
Plate impedance at 45 volts . . .	40,000 ohms.
Amplification factor	30
Recommended grid leak	1 to 2 megohms.

7

DeForest Brochure

DeForest D-01A
DeForest Radio Co., Jersey City, N.J.

Detron D227
Detroit Radio Products Corporation, Detroit, Michigan

Diana D199
Diana Laboratories, New York, N.Y. (Joe Knight Collection)

Diamond D171 A.C.
Diamond Electric Corporation, Newark, N.J.

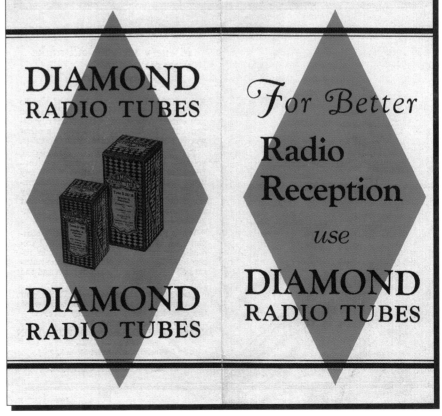

Diamond Brochure

Radio, **December 1929**

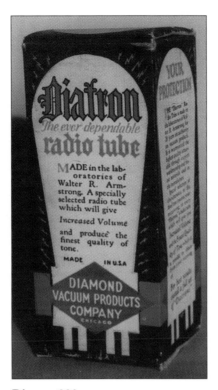

Diatron 222
*Diamond Vacuum Products Co.,
Chicago, Illinois*

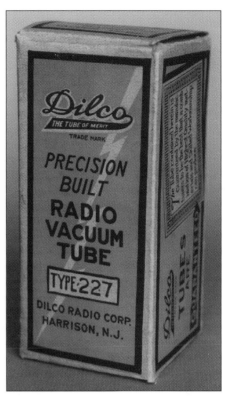

Dilco 227
*Dilco Radio Corporation, Harrison, N.J.
(Clyde Watson Collection)*

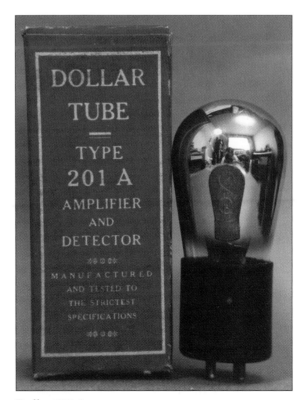

Dollar 201 A
*The Dollar Tube Co., Newark, N.J. (Joe Knight
Collection)*

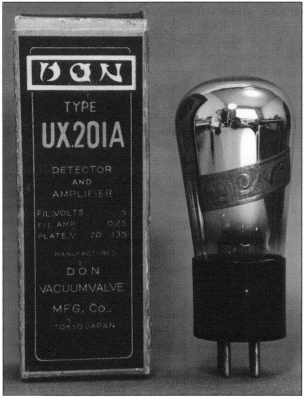

Don UX.201A
Don Vacuumvalve Mfg. Co., Tokyo, Japan (Joe Knight Collection)

Donle-Bristol B-6 Detector, 1926
The Donle-Bristol Corporation, Meriden, Ct. (Joe Knight Collection)

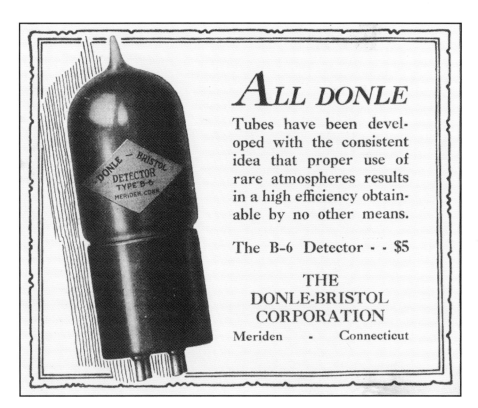

ALL DONLE

Tubes have been developed with the consistent idea that proper use of rare atmospheres results in a high efficiency obtainable by no other means.

The B-6 Detector - - $5

THE
DONLE-BRISTOL
CORPORATION
Meriden   -   Connecticut

Radio News, **October 1926**

Duovac 120
Duovac Radio Tube Corporation, Brooklyn, N.Y. (Clyde Watson Collection)

Duraco UX 200A
Duratron Products Corp., Union City, N.J. (Joe Knight Collection)

Duratron 201-A

Duratron 201-A

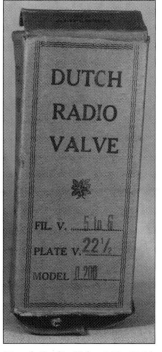

Dutch Radio Valve 200, 1924
*D.R.V. Importing Co.,
Newark, N.J. (Larry Daniel
Collection)*

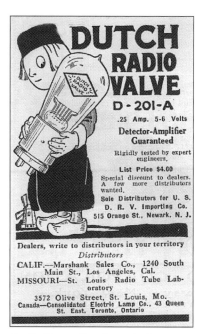

Radio News, October 1924

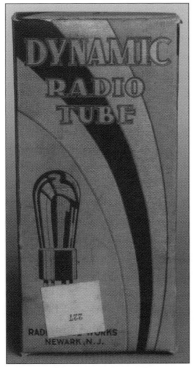

Dynamic 227
Dynamic Radio Tube Works,
Newark, N.J.

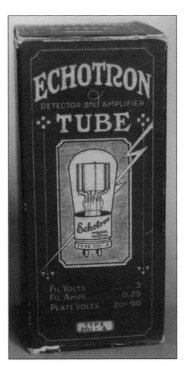

Echotron 201-A

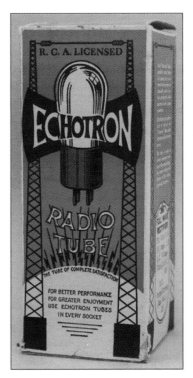

Echotron 280
(Larry Daniel Collection)

Eclipse EX-201-A
Eclipse Tube Works, Newark, N.J.

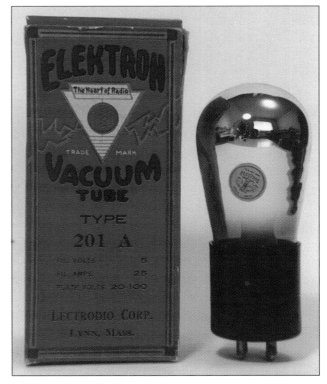

Elektron 201A, 1925
Lectrodio Corporation, Lynn, Mass. (Joe Knight Collection)

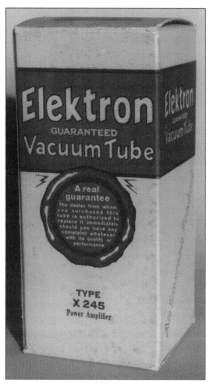

Elektron X 245
Lectrodio Corporation, Lynn, Mass.

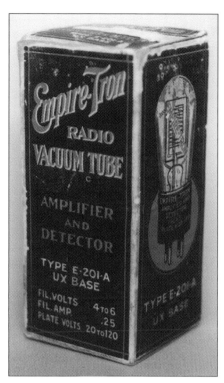

Empire-Tron E-201-A
*Empire Electrical Products Co.,
New York, N.Y.*

Ergon EX-226
*Ergon Electric Corporation, Brooklyn,
N.Y. (Les Rayner Collection)*

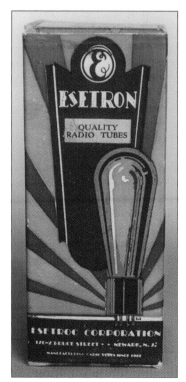

Esetron X201A
*Esetroc Corporation, Newark, N.J.
(Clyde Watson Collection)*

Everbest 227
*Everbest Radio Corp., New York,
N.Y. (Larry Daniel Collection)*

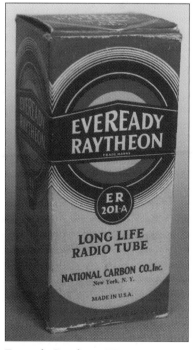

Eveready Raytheon ER 201A, 1929
*National Carbon Co., Inc.,
New York, N.Y.*

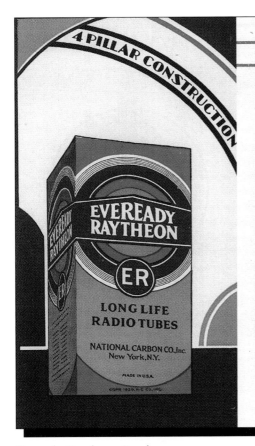

Eveready Raytheon Brochure

EVEREADY RAYTHEON

Radio Tubes

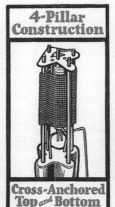

4-Pillar Construction

Cross-Anchored Top and Bottom

EVEREADY RAYTHEON 4-pillar radio tubes are built to give consistently good service over long periods of time. Their performance is uniform day after day and they can be depended on to give maximum results.

These highly desirable qualities are obtained through the use of the unique *"Four Pillar Construction"* (see illustration) in which the elements are cross-anchored at top and bottom. In this way, support is obtained at 8 points instead of 2 as in ordinary tubes.

This rugged trouble-proof feature enables Eveready Raytheon tubes to "stand up" under the most severe handling and shipping conditions and insures the same high degree of performance in the set of the consumer as they exhibit on factory tests.

Engineering vision, painstaking research, and scientific resourcefulness in the Eveready Raytheon Laboratories has each played its part in producing the present high standard of Eveready Raytheon tube efficiency and long life.

There is an Eveready Raytheon tube for every radio purpose, including Television.

"Licensed under patents owned and/or controlled by RCA."

Everest World Top 1-A
Everest Manufacturing Co., Central Falls, R.I. Equivalent to UV-201A.

Everite 171-A
Everite Tube Co., Providence, R.I.

Exceltron 199
Excell Laboratories, New York, N.Y.

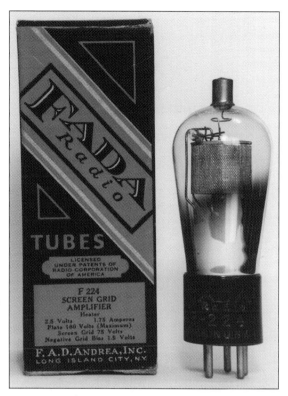

FADA F 224
F.A.D. Andrea, Inc., Long Island City, N.Y. (Joe Knight Collection)

Falck 201-A
Advance Electric Co., Los Angeles, California (Joe Knight Collection)

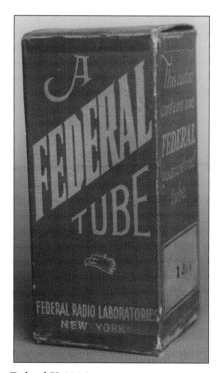

Federal X 201A
Federal Radio Laboratories, New York, N.Y.

Field's 201A
Henry Field Seed Co., Shenandoah, Iowa

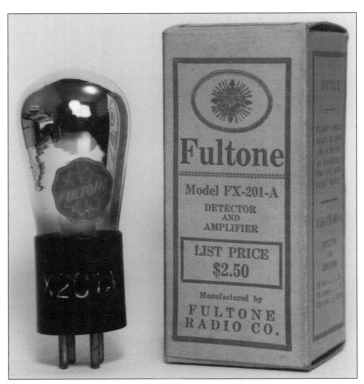

Fultone FX-201-A
Fultone Radio Co. (Joe Knight Collection)

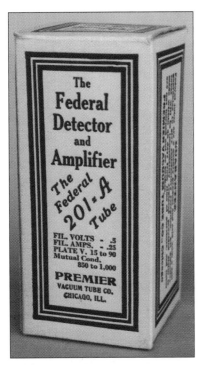

Federal 201-A
Premier Vacuum Tube Co., Chicago, Illinois (Kenneth and Debra Krol Collection)

Gem 201-A, 1925
Gem Tube Co., New York, N.Y. (Joe Knight Collection)

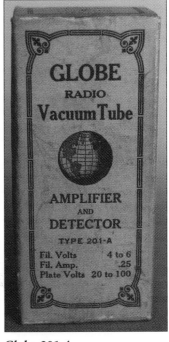

Globe 201-A
Globe Electric Co., Pittsburgh, Pa.

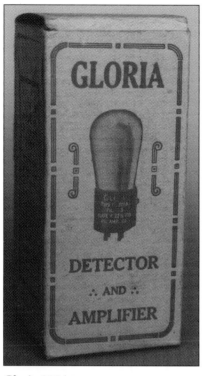

Gloria 201A
Gloria Tube Works, Newark, N.J.

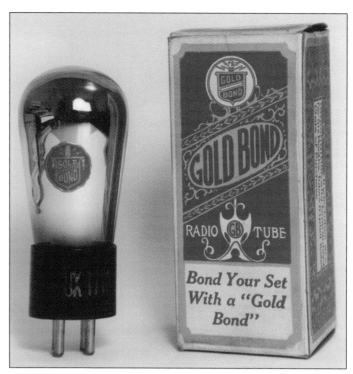

Gold Bond UX 171A
(Joe Knight Collection)

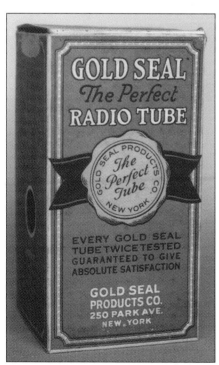

Gold Seal 201A
Gold Seal Products Co., New York, N.Y.

Gold Seal GSX 280
Gold Seal Electrical Co., Inc., New York, N.Y.

Grand Tone X281
F&W Grand, 5-10-25c Stores, Inc.

Gold Seal Radio Tubes

THE SEAL OF PERFECTION

PERFECTION is only attainable by infinite care, continuously exercised, in every smallest detail—PLUS the knowledge and skill of long experience.

Nowhere is this better shown than in the manufacture of radio tubes. Here, incredible deftness must be combined with the finest materials to produce the desired results.

All radio tubes *look* alike—it is performance that makes the distinction between the ordinary tube and PERFECTION.

GOLD SEAL RADIO TUBES demonstrate their perfection by actual performance. A set of Gold Seal tubes is usually a revelation to the radio lover, bringing a new appreciation of what radio reception can be. In full, rounded tones, with all the rich overtones that "make it sound natural," you get the real value of musical instruments and the human speaking or singing voice. Every Gold Seal Tube must pass a rigid test before it leaves the factory. Anything short of perfection causes a tube to be discarded. Gold Seal means something! It has a reputation which must be maintained.

EVERY GOLD SEAL TUBE was perfect when it left the factory but occasionally the delicate parts are damaged in shipment. So every Gold Seal tube is absolutely guaranteed. It must function perfectly under the conditions for which it is designed or it will be replaced without question.

Put a complete set of Gold Seal Radio tubes in your radio receiver and note the improvement in volume, distance, and tone quality. Enjoy PERFECTION.

Type GSX-201a Price $2.00

Type GSX-Hy Mu Price $4.00

Type GSX-112 Price $6.50

Type GSX-120 Price $2.50

Type GSX-199 Price $2.25

Gold Seal Brochure

Green Seal 201-A

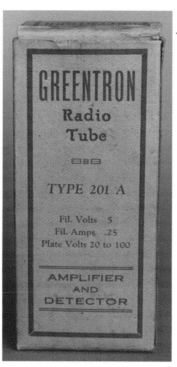

Greentron 201-A

Guaranteed Radio Tube 245
Guaranteed Manufacturing Co.

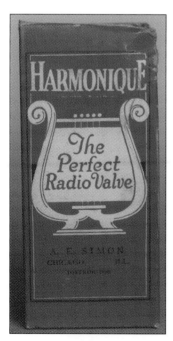

Harmonique 171-A
A.E. Simon, Chicago, Illinois

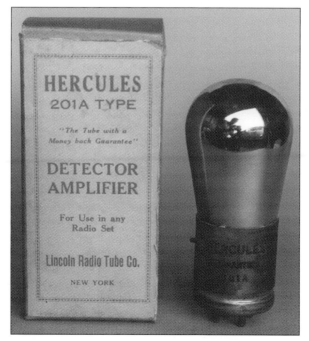

Hercules 201A
Lincoln Radio Tube Co., N.Y.

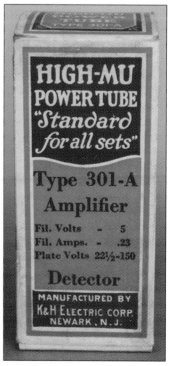

High-Mu 301-A
K&H Electric Corp., Newark, N.J. (Kenneth and Debra Krol Collection)

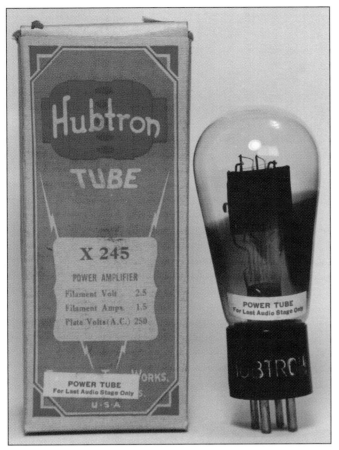

Hubtron X 245

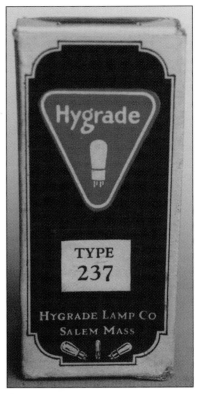

Hygrade 237
Hygrade Lamp Co., Salem, Mass.

Hylon UX-199

Hytron X112-A
Hytron Corporation, Salem, Mass.

Hy-Vac 201A
Hy Vac Radio Co., Newark, N.J.

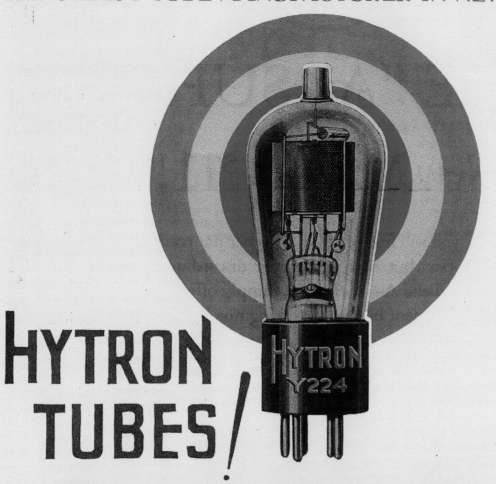

Imperial 227
*Imperial Radio Corp., Chicago,
Illinois*

Jackson 37
Peter Jackson Corporation

JRC JX 201A
*Johnsonburg Radio Corporation,
Johnsonburg, Pa.*

Jewel UX 201-A
*Jewel Radio Co., New York, N.Y.
(Clyde Watson Collection)*

KMA 201-A
*May Seed & Nursery Co., Shenandowa, Iowa (Joe Knight
Collection)*

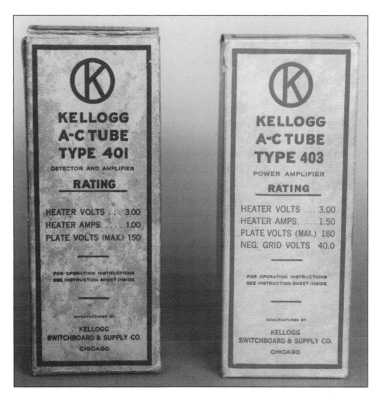

Kellogg 401, 1926
*Kellogg Switchboard & Supply Co., Chicago,
Illinois (Joe Knight Collection)*

Kellogg 401, 403, 1926
Kellogg Switchboard & Supply Co., Chicago, Illinois

KR Amplifier, 1923
*The Ken-Rad Company, Inc., Owensboro, Kentucky (Joe Knight
Collection)*

KR 0201-A
*The Ken-Rad Company, Inc., Owensboro, Kentucky (Joe
Knight Collection)*

Archatron UX 201-A
The Ken-Rad Corporation,
Owensboro, Kentucky

Ken-Rad Archatron UX 201-A
The Ken-Rad Corporation,
Owensboro, Kentucky

Ken-Rad UY-224-A
The Ken-Rad Corporation,
Owensboro, Kentucky

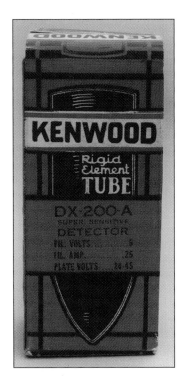

Kenwood DX-200-A
Kenwood Radio Corporation
(Clyde Watson Collection)

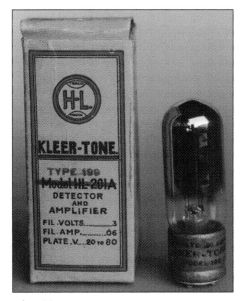

Kleer-Tone 199

Knight Detector
Equivalent to UX200A

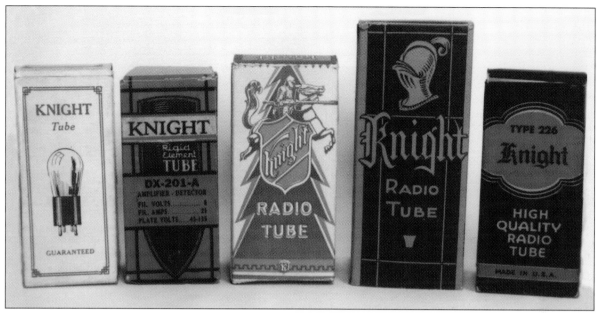

Knight
Assorted Boxes (Joe Knight Collection)

Knight UX-112, 1926
(Joe Knight Collection)

Kwik-lite 201-A
The Usona Manufacturing Co., Inc., Toledo, Ohio (Joe Knight Collection)

LaSalle LS 112A
*LaSalle Products, Chicago,
Illinois (Larry Daniel
Collection)*

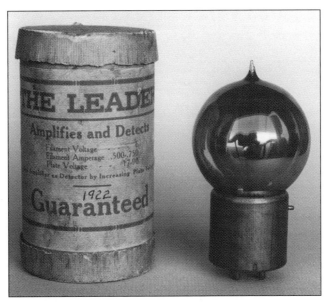

The Leader, Detector and Amplifier

Leader 201-A
*Le Radion Manufacturing Co.,
Chicago, Illinois*

Leader 199-V
*Leader Tube Co., Springfield,
Mass.*

Liberty 81
The Falcon Tube Co., Chicago, Illinois (Joe Knight Collection)

Little Giant 27
*L&G Tube Co., Providence, R.I.
(Kenneth and Debra Krol
Collection)*

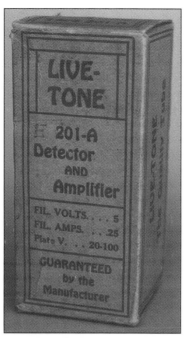

Live-Tone 201-A
(Kenneth and Debra Krol Collection)

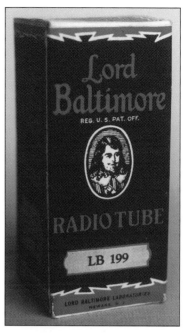

Lord Baltimore LB 199
*Lord Baltimore Laboratories, Newark,
N.J.*

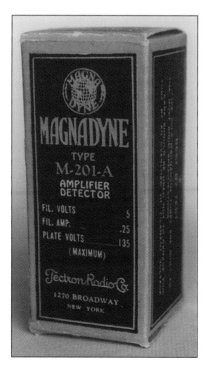

Magnadyne M-201-A
*Tectron Radio Co., New York, N.Y.
(Clyde Watson Collection)*

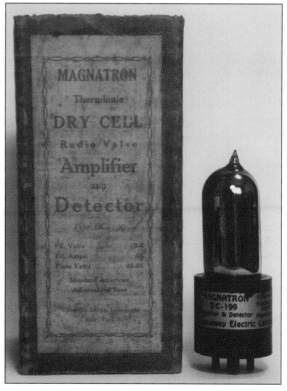

Magnatron DC-199, 1925
*Connewey Electric Laboratories, Hoboken, N.J. (Joe
Knight Collection)*

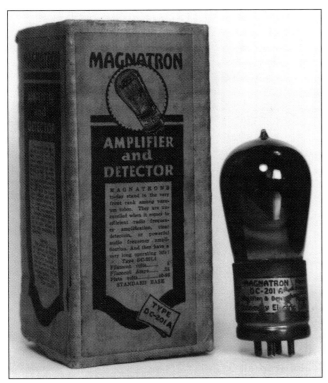

Magnatron DC-201A, 1925
Connewey Electric Laboratories, Hoboken, N.J. (Joe Knight Collection)

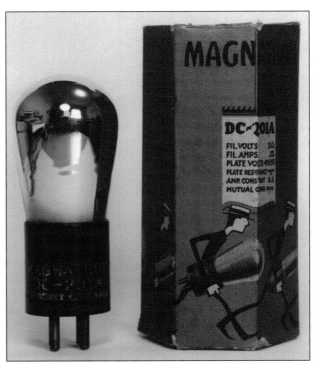

Magnatron DC-201A
Connewey Electric Laboratories, Hoboken, N.J.

Magnavox A, 1924
The Magnavox Co., Oakland, California. This was advertised as a "gridless" tube, no doubt an attempt to circumvent the DeForest "interposed grid" patent. Instead it employed a "control electrode." (Joe Knight Collection)

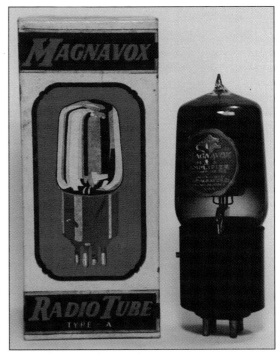

Magnavox A, 1924
The Magnavox Co., Oakland, California (Joe Knight Collection)

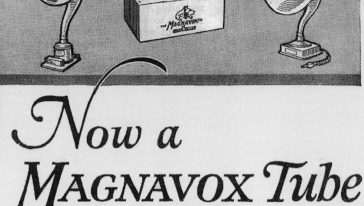

Now a MAGNAVOX Tube

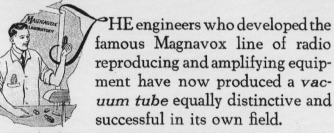

THE engineers who developed the famous Magnavox line of radio reproducing and amplifying equipment have now produced a *vacuum tube* equally distinctive and successful in its own field.

One trial convinces the most exacting user that the Magnavox will replace ordinary tubes to great advantage in any receiving set.

Magnavox Products

Reproducers of electro-dynamic and semi-dynamic type, for all vacuum tube receiving sets;
$25.00 to $50.00

Power Amplifiers for audio-frequency amplification, one, two, and three-stage;
$27.50 to $60.00

Combination Sets combining a Reproducer and Power Amplifier in one unit;
$59.00, $85.00

Vacuum Tubes: A storage battery tube of new and approved design for all standard circuits . . . $5.00

Magnavox Radio Products are sold by reliable dealers everywhere. If unacquainted with the Magnavox store in your vicinity, write us for information.

The name Magnavox is your assurance of quality and efficiency.

THE MAGNAVOX CO., OAKLAND, CALIF.
NEW YORK   SAN FRANCISCO
Canadian Distributors: Perkins Electric Limited, Toronto, Montreal, Winnipeg

$5.00

MAGNAVOX RADIO VACUUM TUBE TYPE A is a storage battery tube for use as audio frequency and radio frequency amplifier in all standard circuits. Highly recommended also for detector use. This tube is not critical of adjustment either as to plate or filament. Filament consumption one quarter of an ampere.

The most notable feature of the new Magnavox Radio Tube consists in eliminating the grid.

Unlike the ordinary storage battery tube, Magnavox Tubes give the electrons an unobstructed passage between filament and plate, with the result that the Magnavox has less than one half the internal capacity of other tubes of similar type.

10 R

Magnavox 171
*The Magnavox Co.,
Oakland, California
(Clyde Watson Collection)*

Majestic Full Wave Rectifier
Grigsby-Grunow-Hinds Co., Chicago, Illinois

Majestic G-45
*Grigsby-Grunow Co., Chicago,
Illinois (Clyde Watson Collection)*

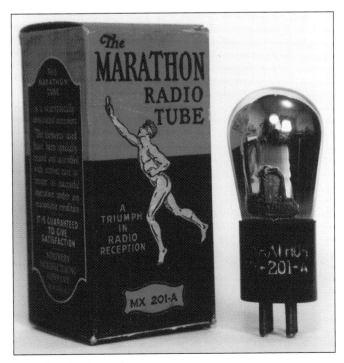

Marathon MX 201-A
Northern Manufacturing Co., Newark, N.J.

Marconi/Moorhead/DeForest VT Amplifier, 1919
*Marconi Wireless Telegraph Co. of America, New York, N.Y.
(Joe Knight Collection)*

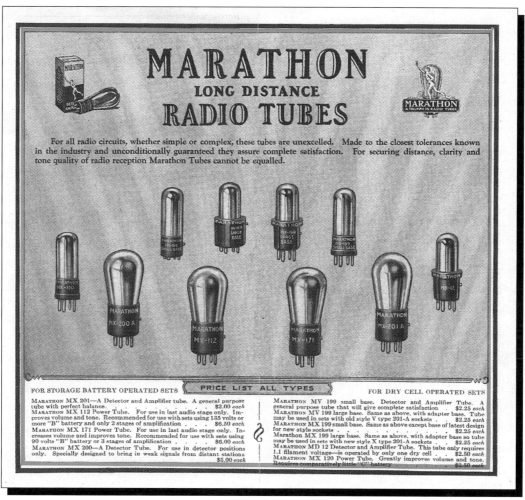

Marathon Brochure

Marvin RY401
Marvin Radio Tube Corporation, Irvington, N.J. Filament wire is on outside of glass.

Master Sodium Detector
Electro Chemical Laboratories, New York, San Francisco (Joe Knight Collection)

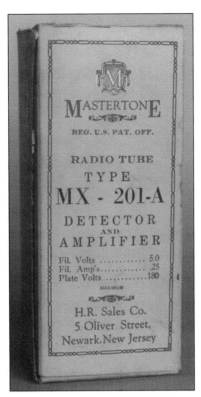

Mastertone MX-201-A, 1925
H.R. Sales Co., Newark, N.J.

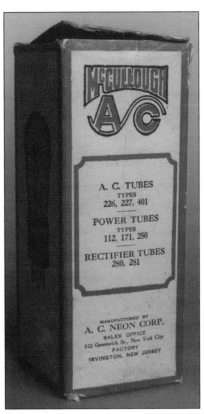

McCullough 227
A.C. Neon Corporation, Irvington, N.J.

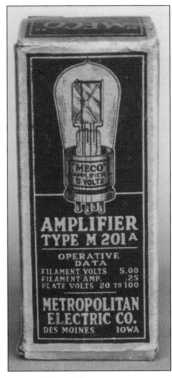

Meco M 201A
Metropolitan Electric Co., Des Moines, Iowa (Kenneth and Debra Krol Collection)

Mello-Tron X-200-A
Mellotron Tube Co., Chicago, Illinois (Larry Daniel Collection)

Melophonic UX-201-A

Meteor 01A

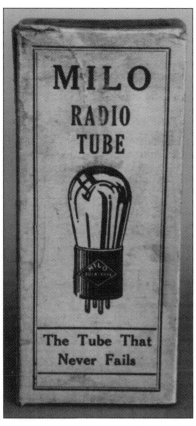

Milo X-250
(Kenneth and Debra Krol Collection)

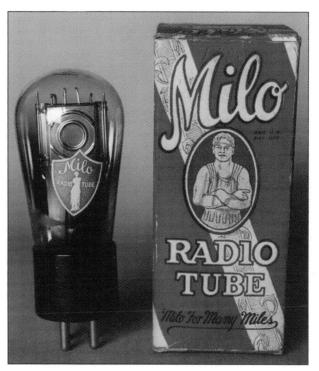

Milo UX-112

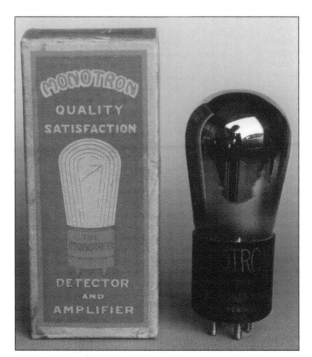

Monotron 201A
Paxton and Gallagher Co., Omaha, Nebraska

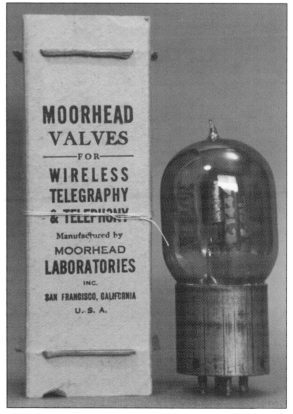

Moorhead SE-1444, 1918
Moorhead Laboratories Inc., San Francisco, California.
WWI U.S. Navy type. (Joe Knight Collection)

Music Master A
Music Master Corporation, Philadelphia, Penn.
Equivalent to 201A. These had green colored glass. (Joe
Knight Collection)

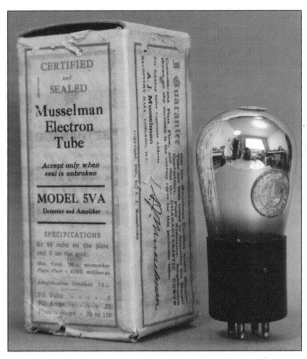

Musselman 5VA
A.J. Musselman, Inc., Chicago, Illinois. These were
manufactured for Musselman by the Van Horne Co. of
Franklin, Ohio. (Joe Knight Collection)

Musselman 5VA
A.J. Musselman, Inc., Chicago, Illinois (Joe Knight Collection)

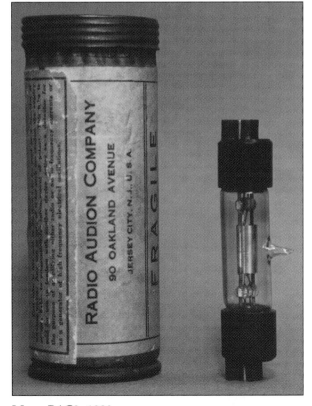

Myers RAC3, 1920
Radio Audion Company, Jersey City, N.J. Shown with
mailing tube. (Joe Knight Collection)

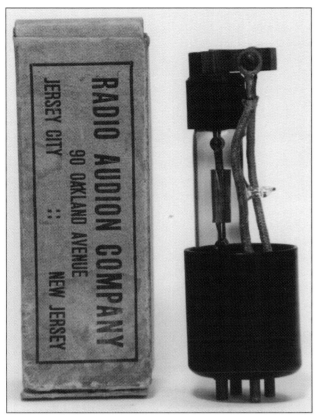

Myers RAC3, 1920
Radio Audion Company, Jersey City, N.J. Shown in UV base adaptor. (Joe Knight Collection)

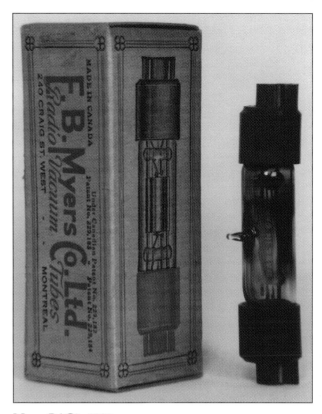

Myers RAC3, 1923
E.B. Myers Co. Ltd., Montreal, Canada. The company moved to Canada to avoid RCA patent infringement battle. (Joe Knight Collection)

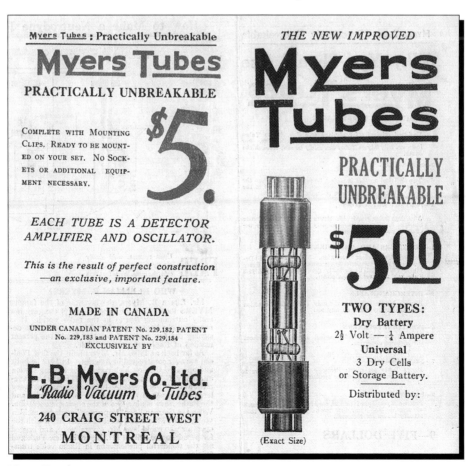

Myers Brochure

Radio News, March 1926

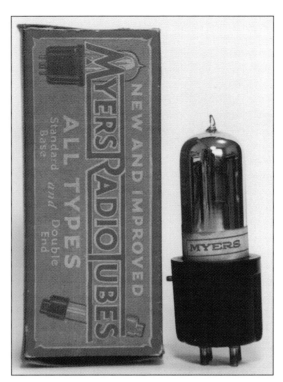

Myers 99, 1926
Myers Radio tube Corporation, Cleveland, Ohio (Joe Knight Collection)

National Union NY224
National Union Radio Corporation, New York, N.Y.

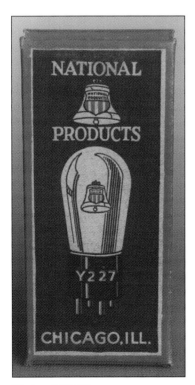

National Y227
National Products, Chicago, Illinois

Neon 245
Guarantee Tube Co., New York (Kenneth and Debra Krol Collection)

Neontron 201A
Neontron Co., New York, N.Y.

Neptron X200-A

List $4.00

Supersensitive Detector Tube

The NEPTRON X-200-A detector tube is of the caesium vapor, supersensitive type, and is used in receiving sets operating from a storage battery where especially sensitive conditions are desired for D X work.

The filament is operated at five volts, preferably controlled by a 20-ohm rheostat; a plate voltage of 45 volts is employed. The grid-return lead is connected to the plus side of the filament, and a grid leak of two to three megohms is recommended. A grid condenser having a capacity of .00025 microfarads is required.

The NEPTRON X-200-A is not made for use other than as detector for weak signals. It usually requires one to three minutes after turning on for the tube to become stabilized.

Neptron X-199

List $2.25

Dry-Cell Receiving Tube

The NEPTRON X-199 tube is a general purpose tube similar in its characteristics to the X-201-A, but of smaller power and designed especially for operation from dry cell batteries.

The NEPTRON X-199 tube is suitable for operation as a radio frequency amplifier detector, or audio frequency amplifier.

The filament is operated at 3.3 volts and draws only 1/16 ampere. The plate and grid voltages, with grid leak and condenser values, are the same as employed for the X-201-A.

Neptron X201-A

List $1.50

General Purpose Detector and Amplifier

The NEPTRON X-201-A is the most generally used D. C. tube at the present time, being equally satisfactory as a radio frequency amplifier, detector or audio frequency amplifier.

The filament is of the thoriated tungsten type and is usually operated from a six-volt storage battery. The current consumption is .25 ampere at 5 volts. Plate voltages of from 45, when used as a detector, to 135, when used as last stage audio amplifier, are used.

Neptron Corporation
Salem Mass

Neptron Brochure

Neptron X 201-A
Neptron Corporation, Beverly, Mass.

Neptune X. 199
Neptune Tube Works, Newark, N.J.

OK OKX-201-A
(Clyde Watson Collection)

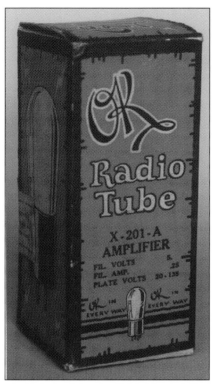

OK X-201-A

OK 01A

Odeon AC-226
Odeon Manufacturing Co.,
Newark, N.J.

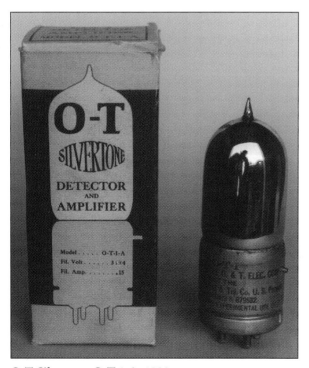

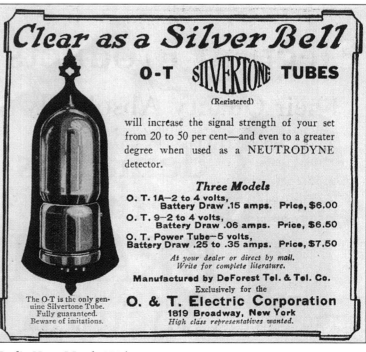

O-T Silvertone O-T-1-A, 1923
O.&T. Electric Corporation, New York, N.Y.

Radio News, **March 1924**

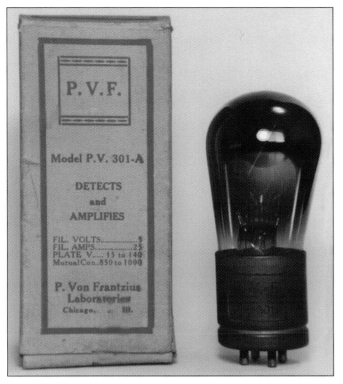

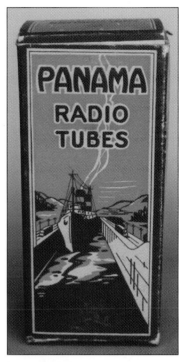

P.V.F. P.V.301-A, 1925
P. Von Frantzius Laboratories, Chicago (Joe Knight Collection)

The Pace Setter 12A
Dean Phipps Auto Stores, Pa., N.Y., N.J.

Panama 01A
(Kenneth and Debra Krol Collection)

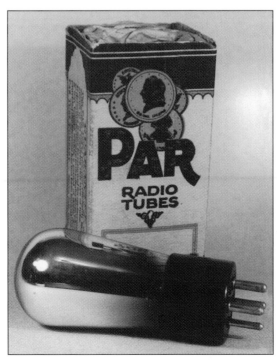

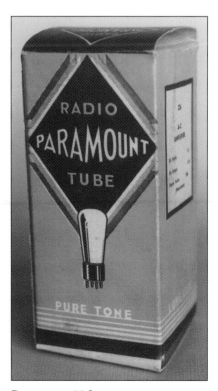

Panama 01A

PAR UX-201A
PAR Tube Co., Newton Falls, Ohio (Larry Daniel Collection)

Paramount 226
Paramount Radio Tube Co., Brooklyn, N.Y.

Peak 210A
(Clyde Watson Collection)

Perfection 199
Perfection Tube Co.,
Chicago, Illinois

Perfectone 210
(Joe Knight Collection)

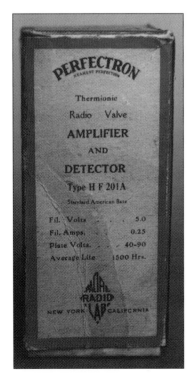

Perfectron HF 201A
Pacific Radio Lab, Los Angeles,
California

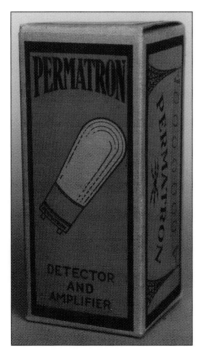

Permatron UX 201A
Permatron Tube Co., Union City, N.J.

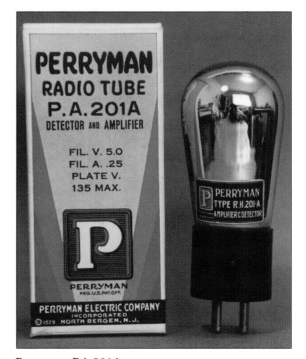

Perryman P.A.201A
Perryman Electric Co., Inc., North Bergen, N.J. (Joe
Knight Collection)

LIKE TWO VISES

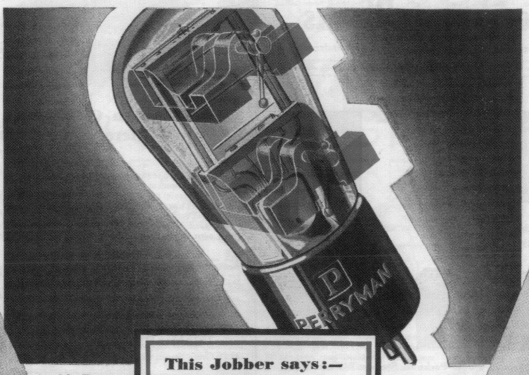

This Jobber says:—

" . . . and Perryman is the only brand of tube we have ever carried on which we did not have a loss. We have found that Perryman gives much better satisfaction, fewer service calls, resulting in more satisfied dealers, and naturally more satisfaction to us."

The double Perryman Bridge grips the elements in Perryman Tubes, top and bottom. It holds the grid, plate and filament in permanent parallel alignment. This absolutely assures uniform operation of every Perryman Tube.

With this sturdy bridge construction Perryman Tubes defy all necessary handling in shipment, in your store and in your customers' sets.

The Tension-Spring, another exclusive Perryman feature, allows for the uniform expansion and contraction of the filament due to temperature changes.

Both these features mean fewer replacements—greater net profits to you.

Point out the double Perryman Bridge and Tension-Spring to your customers

THE PERRYMAN ELECTRIC CO., INC., 4901 Hudson Blvd., North Bergen, N. J.

PERRYMAN
RADIO [P] TUBES

Perryman R.H.201-A, R.H.201-A, P.A.201A
Perryman Electric Co., Inc., North Bergen, N.J.

Philmore 124
Philmore Manufacturing Co., Inc., N.Y.

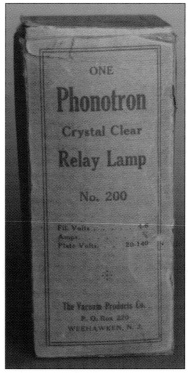

Philco 01A
Philco Philadelphia, Pa. (Joe Knight Collection)

Phonotron 200
The Vacuum Products Co., Weehawken, N.J. (Clyde Watson Collection)

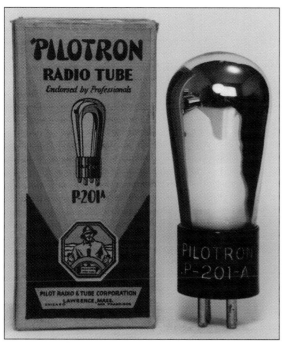

Pilotron P-201A
Pilot Radio & Tube Corporation, Lawrence, Mass. (Joe Knight Collection)

Pingree 201A
Pingree Radio Service, Boston, Mass. (Larry Daniel Collection)

Pioneer 201-A

Platron S 201-A
Platron Radio Co., New York, N.Y.

Premier X-201A
Wholesale Radio Supply Co., San Francisco, California

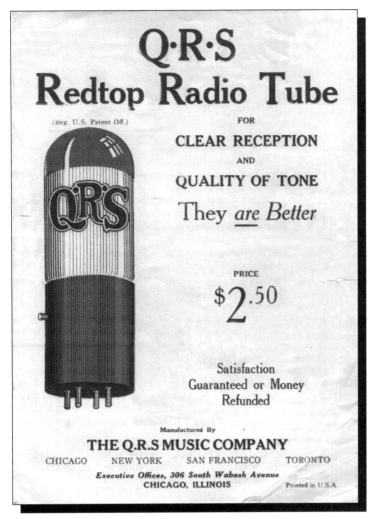

QRS Brochure

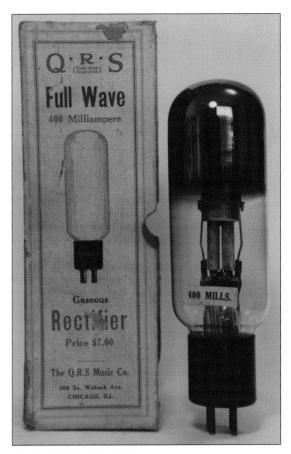

QRS Full Wave Rectifier, 1927
The QRS Music Company, Chicago, Illinois (Joe Knight Collection)

QRS Full Wave Rectifier, 1927
The QRS Music Company, Chicago, Illinois

QRS Redtop Power Tube
The QRS Music Company, Chicago, Illinois

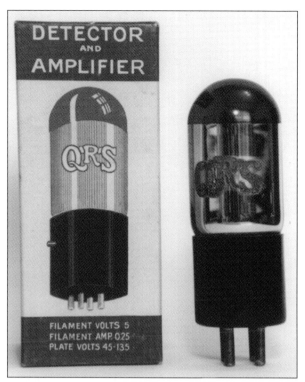

QRS Redtop Detector & Amplifier
The QRS Music Company, Chicago, Illinois (Joe Knight Collection)

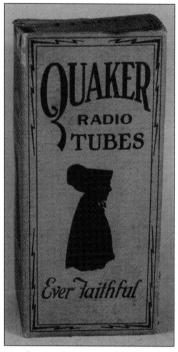

Quaker 201A
Gardiner & Hepburn, Philadelphia (Larry Daniel Collection)

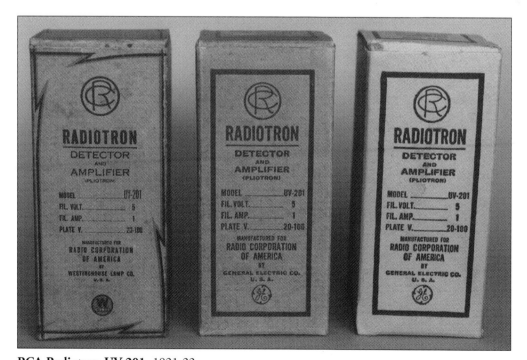

RCA Radiotron UV-201, 1921-22
Manufactured for Radio Corporation of America by Westinghouse Lamp Co.(left box) and General Electric Co.(middle and right boxes)

RADIO CORPORATION of AMERICA

RADIOTRON U. V. 201

A NEW AMPLIFIER OF THE PLIOTRON TYPE
FOR
AMATEUR AND EXPERIMENTAL WIRELESS STATIONS

COMPLEX amplifying circuits for the magnification of radio and tone frequency currents require an amplifying Vacuum Tube of *rigid* operating characteristics. There is an increasing demand among radio experimenters for a vacuum tube amplifier which will magnify the telephone currents in a radio receiving set and which can be shifted from one socket to another in a cascade outfit without loss of signal audibility. Moreover, the amplifier must be free from the tube "noises" accompanying the use of improperly designed vacuum tubes.

RADIOTRON U. V. 201, the second of the new series of Vacuum Tubes designed by the engineers of the Research Laboratory of the General Electric Company for the Radio Corporation, possesses the qualifications outlined above and it should be a part of every experimental radio receiving station. U. V. 201 may be used as a *detector*, or as a *tone frequency* or *radio frequency amplifier*.

In cascade radio frequency amplifying circuits, U. V. 201 can be adjusted to magnify without distortion. The use of such circuits is on the increase in amateur stations, particularly where long distance communication is desired on short wave lengths (200 meters or less).

As a *detector* the best results are secured from *Radiotron U. V. 201* with a *grid condenser* of approximately .0001 mfd. capacity and with a shunt GRID LEAK of ½ to 2 megohms, according to the type of circuit employed.

The normal plate voltage of Radiotron U. V. 201 is 40 volts, although increasing amplifications can be obtained at plate voltages up to 100. At 40 volts on the plate, the amplification constant varies from 6.5 to 8; at 100 volts on the plate, from 8 to 10. The output impedance varies from 15,000 ohms to 25,000 ohms at 40 volts on the plate, and from 10,000 to 15,000 with 100 volts on the plate.

The normal *filament current* for RADIOTRON U. V. 201 is approximately 1 ampere. The filament is designed for connection to the terminals of a 6-volt storage battery with a standard filament rheostat in series.

To obtain maximum amplification with U. V. 201 means should be supplied for placing negative potentials on the grid, although good amplification may be secured without any special provision for such potentials. The requisite negative grid potential for the use of U. V. 201 in *amplification circuits* can be secured by connecting a standard "C" battery of two or three volts in the grid circuit, shunted by a 200 to 400 ohm potentiometer, or by placing a 2 ohm resistance in series with the negative terminal of the filament and connecting the "low" potential" terminal of the tuner secondary to include this resistance in the grid circuit. The latter method usually provides the requisite grid potential results, but the proper value for maximum amplification is generally best found by trial and experiment, with a variable source of e.m.f. supplied locally.

IMPORTANT FACTS CONCERNING RADIOTRON U. V. 201

The Radio Corporation's gas content detector and amplifier tube, RADIOTRON U. V. 200, is in itself an excellent tone frequency amplifier, but it does not give the "power" amplifications obtainable from the Corporation's SPECIAL AMPLIFIER TUBE RADIOTRON U. V. 201. Thus, for devices requiring a considerable amount of energy for their

1920

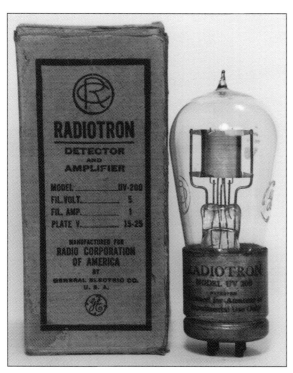

RCA Radiotron UV-200, 1921
General Electric Co., Schenectady, N.Y. (Joe Knight Collection)

RCA Aeriotron WR-21, 1922
Westinghouse Electric & Mfg. Co., East Pittsburgh, Pa. Used only in the Westinghouse "Aeriola Grand" receiver. (Joe Knight Collection)

Westinghouse Aeriotron WR-21, 1922
Westinghouse Electric & Mfg. Co., East Pittsburgh, Pa. Same as above, only packaged under the Westinghouse name. (Joe Knight Collection)

Westinghouse WB-800, 1922
Westinghouse Electric & Mfg. Co. East Pittsburgh, Pa. Ballast tube for use in the "Aeriola Grand". (Joe Knight Collection)

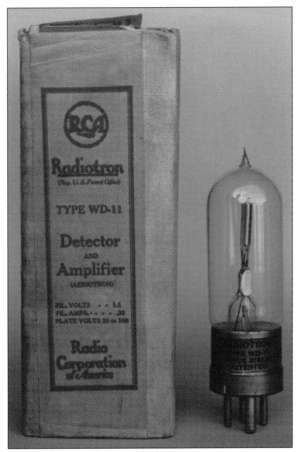

RCA Radiotron WD-11 (Aeriotron), 1923
Westinghouse Electric & Mfg. Co., East Pittsburgh, Pa. (Les Rayner Collection)

RCA Radiotron WD-12, 1923
Westinghouse Electric & Mfg. Co., East Pittsburgh, Pa. The WD-12 is the same electrically as the WD-11, but with a UV-type base.

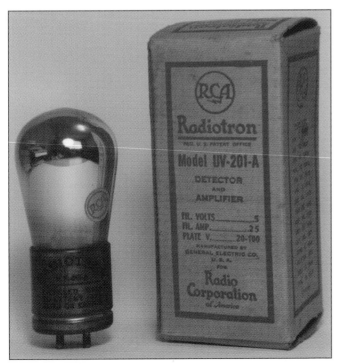

RCA Radiotron UV-201-A, 1923
General Electric Co., Schenectady, N.Y. (Joe Knight Collection)

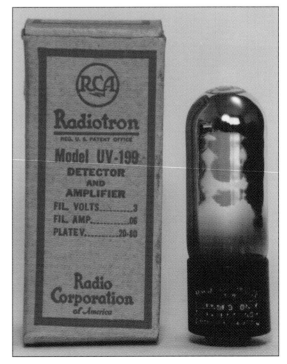

RCA Radiotron UV-199, 1924
Radio Corporation of America, New York, N.Y. (Joe Knight Collection)

RCA Radiotron UX-201-A, 1925
Radio Corporation of America, New York, N.Y. (Joe Knight Collection)

RCA Rectron 213, 1925
Radio Corporation of America, New York, N.Y. (Joe Knight Collection)

RCA Brochure

RCA UX-245, 1929
Radio Corporation of America, New York, N.Y. (Joe Knight Collection)

RCA Victor 36
RCA Manufacturing Company Inc., Camden, N.J. (Clyde Watson Collection)

RA-DEX C
(Kenneth and Debra Krol Collection)

Raytheon B, 1925
Raytheon Manufacturing Co., Cambridge, Ma. Full Wave rectifier for use in battery eliminators. (Joe Knight Collection)

Raytheon BA, 1926
Raytheon Manufacturing Co., Cambridge, Ma. Full-Wave rectifier for use in battery eliminators. (Joe Knight Collection)

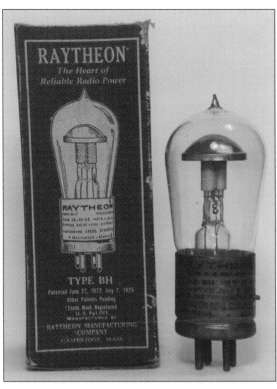

Raytheon BH, 1926
Raytheon Manufacturing Co., Cambridge, Ma. (Joe Knight Collection)

Raytheon BH, 1927
Raytheon Manufacturing Co., Cambridge, Mass. (Joe Knight Collection)

Raytheon 227
Raytheon Manufacturing Co., Cambridge, Mass.

Rectobulb R-81
National Radio Tube Co., San Francisco, California

Regal 201-A
Regal Products Corporation, Philadelphia, Pa.

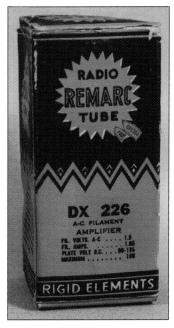

Remarc DX 226
(Larry Daniel Collection)

Rextron UX199
K. & H. Electric Corporation, Newark, N.J.

Roice A, 1925
Roice Tube Co., Newark, N.J. This company abruptly changed their name from "Rolls Royce" to "Roice," as the accompanying magazine ads attest. (Joe Knight Collection)

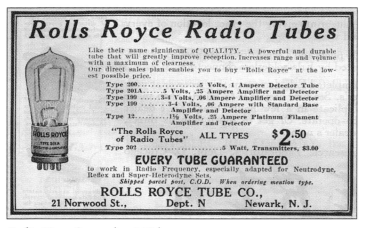

Radio News, **September 1924**

Radio News, **April 1925**

Royalfone 201A, 1925
Royal Electrical Laboratories, Newark, N.J. (Joe Knight Collection)

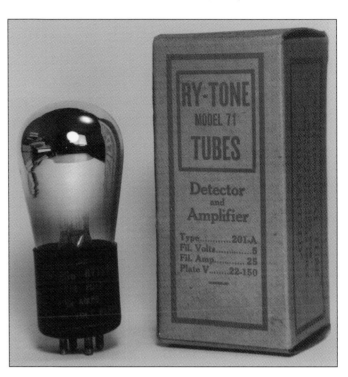

Ry-Tone 201-A
(Joe Knight Collection)

Roxy UX 201
(Larry Daniel Collection)

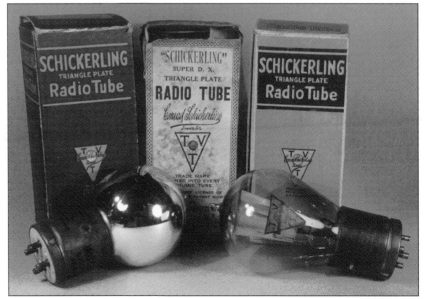

Schickerling S-400 and Z-50, 1925
Schickerling Products Corporation, Newark, N.J. Cartons shown (from left) are for S-2500, SX-201A and S-4000. (Larry Daniel Collection)

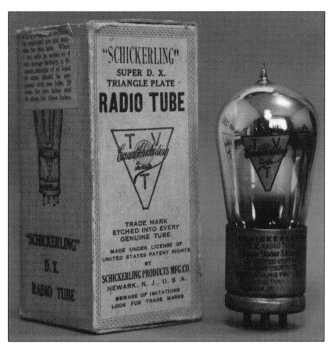

Schickerling S4000, 1925
Schickerling Products Corporation, Newark, N.J.

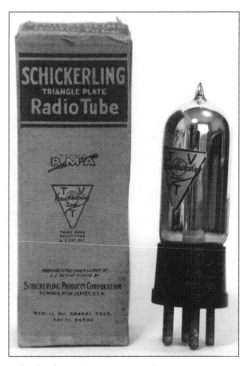

Schickerling SX-8100, 1926
Schickerling Products Corporation, Newark, N.J.
(Joe Knight Collection)

Schickerling Brochure

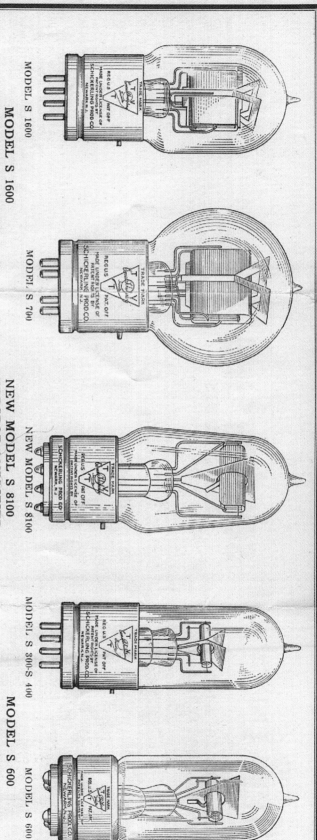

MODEL S 1600

MODEL S 1600
DETECTOR-AMPLIFIER
For Dry Cells or Storage Battery
Price, $4.00

A 5 volt standard base tube, drawing only 16/100 amperes per hour. This tube is designed to operate on a 6 volt storage battery with a 20 ohm rheostat. It also can be operated with three or four No. 6 dry cells in series. As amplifier use plate voltage 45 volts to 90 volts. This tube is likewise a fine detector tube on 22½ to 45 volts. Can be used in any set built for ¼ amp. tubes.

MODEL S 700
POWER TUBE OSCILLATOR AND SUPER AMPLIFIER-DETECTOR
Price, $7.00

5 volt, ¼ ampere power tube. This tube is made to be used in loud speaker sets of any standard make. Clear as a bell. It gives a deep beautiful volume. It reproduces and amplifies without distortion. 90 to 160 plate voltage. It is also used as oscillator in super heterodyne and other sets with splendid success. Use 20 ohm rheostat. No loud speaker set complete without this tube. This is a large round tube—standard base. A tube you will not be without when once used.

MODEL S 700

NEW MODEL S 8100
DETECTOR-AMPLIFIER
Price, $3.00

A 3 to 4 volt 7/100 ampere, miniature base tube. Fits standard 199 sockets. This tube is ideal for portable sets and sets equipped with 199 Type sockets. The plate voltage is 22½-45 volts as detector and 90 volts as amplifier. A 30 or 50 ohm rheostat should be used for operation. Use with three Dry Cells on 3 to 4 volts.

This tube embodies in itself a great improvement over all other miniature dry cell tubes of the 199 type. Its construction eliminates the universally known faults of such tubes. Here we use for the first time the newly discovered radium glass, an invention of Conrad Schickerling (patents pending). This glass of greenish irridescent color gives through its radium activity a most remarkable result in radio tube construction.

NEW MODEL S 8100

MODEL S 300
Price, $3.00
1½ Volt Dry Cell Tube, ¼ Amp.
Amplifier Detector. Standard Base

This tube is recommended for one Dry Cell use. It is a splendid Detector on 22 to 45 volts and Amplifies on 45 to 90 volts. Clear, beautiful tone reproduction are the outstanding features of this tube.

MODEL S 400
3 Volt Dry Cell, 1/10 Amp.
Amplifier Detector. Standard Base
Price, $3.00

Works well on three Dry Cells. It consumes only one-tenth of an ampere. This is a very popular tube. Detector 22 to 45 volt, amplifier 45 to 90 volt. Use with 30 ohm rheostat.

MODEL S 300-S 400

MODEL S 600
DETECTOR-AMPLIFIER
Price, $3.00

A 3 to 4 volt 8/100 ampere, miniature base tube. Fits standard 199 sockets. This tube is ideal for portable sets. The plate voltage is 22½-45 volts as detector and 90 volts as amplifier. A 50 ohm rheostat should be used for operation. Use with three Dry Cells.

MODEL S 200—DETECTOR
Price, $3.00

5 volt, ¼ ampere super detector. Standard base. A very sensitive gas filled tube consuming only ¼ ampere, 16 to 45 plate voltage. The ideal detector tube for any type set. Use 20 ohm rheostat.

MODEL S 600

Schickerling Brochure

Schickerling S-100
Schickerling Products Mfg. Co., Newark, N.J. (Joe Knight Collection)

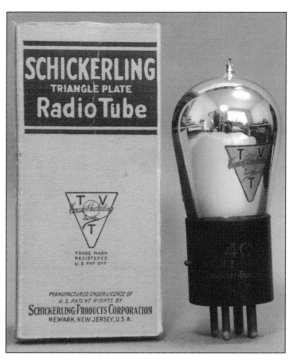

Schickerling SX-400, 1926
Schickerling Products Corporation, Newark, N.J. (Joe Knight Collection)

Shelby 201-A
The Shelby Radio Tube Co., Shelby, Ohio

Shieldplate SP 122, 1928
Shieldplate Tube Corporation, Chicago, Illinois

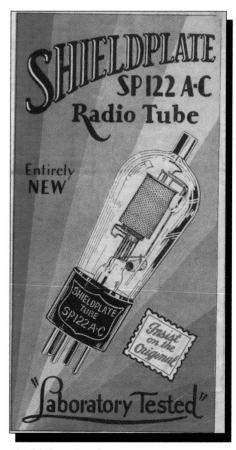

Shieldplate Brochure

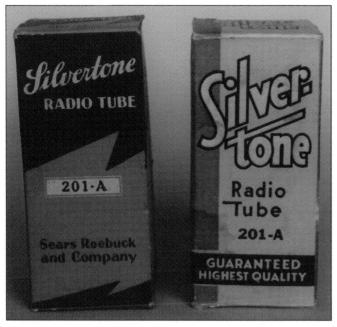

Silvertone 201-A
Sears Roebuck and Company

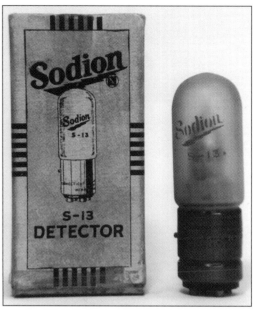

Sodion S-13 Detector, 1923
*Connecticut Telephone & Electric Co.,
Meriden, Ct. (Joe Knight Collection)*

Sodion D-21, 1923
*Connecticut Telephone & Electric Co., Meriden, Ct. (Joe
Knight Collection)*

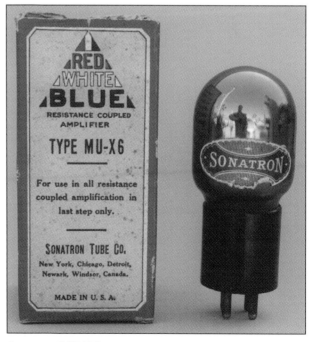

Sonatron MU-X6
*Sonatron Tube Co., Chicago, Illinois (Martin Bergan
Collection)*

Sonatron MU-X30, MU-X20, MU-X6, 1926
Sonatron Tube Co., Chicago, Illinois. "Red, White, Blue" tubes shown on Sonatron resistance-coupled amplifier. (Martin Bergan Collection)

Sonatron 199X
Sonatron Tube Co., Chicago, Illinois

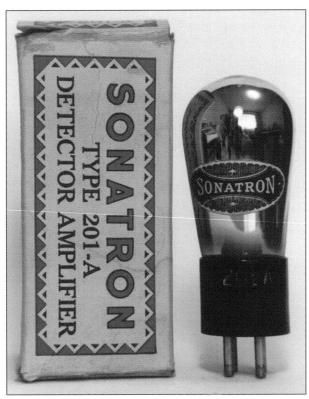

Sonatron 201-A
Sonatron Tube Co., Chicago, Illinois (Joe Knight Collection)

Sonatron X 250
Sonatron Tube Co., Chicago, Illinois

Citizens Radio Call Book, Spring 1928

Songbird X 201-A

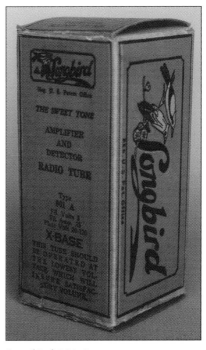

Songbird 901 A

Songbird GX 201A

Songbird 01A
W.T. Grant Co.

Songbird 47
W.T. Grant Co. (Clyde Watson Collection)

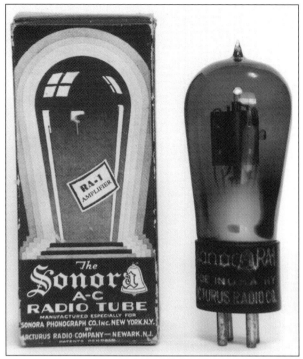

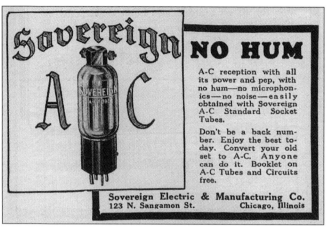

Sonora RA-1, 1928
Sonora Phonograph Co., Inc., New York, N.Y. Non-standard,
15 V. carbon heater. (Joe Knight Collection)

***Citizens Radio Call Book*, Spring 1928**

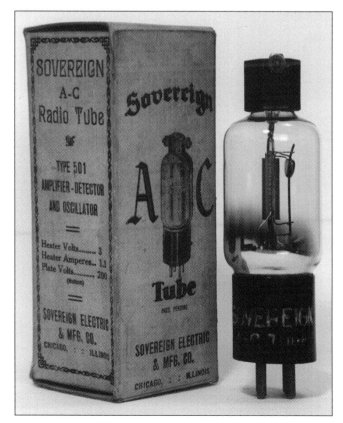

Sovereign AC 501, 1927
Sovereign Electric & Manufacturing Co., Chicago, Illinois (Joe
Knight Collection)

Sparton 485
The Sparks-Withington Co., Jackson,
Michigan

Sparton 401
The Sparks-Withington Co., Jackson, Michigan

Sparton 181
The Sparks-Withington Co., Jackson, Michigan

Speed 201A
Speed Radio Corporation, Long Island City, N.Y.

Speed 12
Cable Supply Co., Inc., New York (Kenneth and Debra Krol Collection)

Speed Triple-Twin 295, 1932
Cable Radio Tube Corp., Brooklyn, N.Y. This was a "2 in 1" tube with an indirectly heated driver triode directly coupled to a directly heated output triode. (Joe Knight Collection)

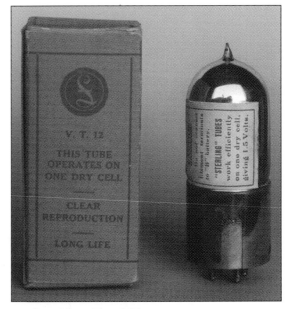

Sterling Silver Tone V.T. 12

Stewart-Warner S-W 501-A, 1925
*Stewart-Warner Speedometer
Corporation, Chicago, Illinois*

Sturdy 201-A
Sturdy Engineering Co., Inc., Chicago, Illinois

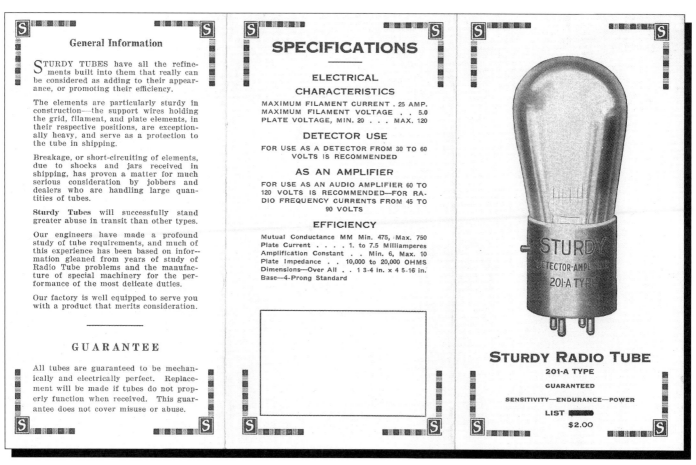

General Information

STURDY TUBES have all the refinements built into them that really can be considered as adding to their appearance, or promoting their efficiency.

The elements are particularly sturdy in construction—the support wires holding the grid, filament, and plate elements, in their respective positions, are exceptionally heavy, and serve as a protection to the tube in shipping.

Breakage, or short-circuiting of elements, due to shocks and jars received in shipping, has proven a matter for much serious consideration by jobbers and dealers who are handling large quantities of tubes.

Sturdy Tubes will successfully stand greater abuse in transit than other types.

Our engineers have made a profound study of tube requirements, and much of this experience has been based on information gleaned from years of study of Radio Tube problems and the manufacture of special machinery for the performance of the most delicate duties.

Our factory is well equipped to serve you with a product that merits consideration.

GUARANTEE

All tubes are guaranteed to be mechanically and electrically perfect. Replacement will be made if tubes do not properly function when received. This guarantee does not cover misuse or abuse.

SPECIFICATIONS

ELECTRICAL CHARACTERISTICS

MAXIMUM FILAMENT CURRENT . 25 AMP.
MAXIMUM FILAMENT VOLTAGE . . 5.0
PLATE VOLTAGE, MIN. 20 . . . MAX. 120

DETECTOR USE

FOR USE AS A DETECTOR FROM 30 TO 60 VOLTS IS RECOMMENDED

AS AN AMPLIFIER

FOR USE AS AN AUDIO AMPLIFIER 60 TO 120 VOLTS IS RECOMMENDED—FOR RADIO FREQUENCY CURRENTS FROM 45 TO 90 VOLTS

EFFICIENCY

Mutual Conductance MM Min. 475, Max. 750
Plate Current . . . . 1. to 7.5 Milliamperes
Amplification Constant . . Min. 6, Max. 10
Plate Impedance . . 10,000 to 20,000 OHMS
Dimensions—Over All . . 1 3-4 in. x 4 5-16 in.
Base—4-Prong Standard

STURDY RADIO TUBE
201-A TYPE
GUARANTEED
SENSITIVITY—ENDURANCE—POWER
LIST
$2.00

Sturdy Brochure

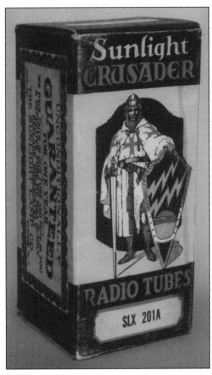

Sunlight Crusader SLX 201A
The Sunlight Lamp Co., Newark, N.J.

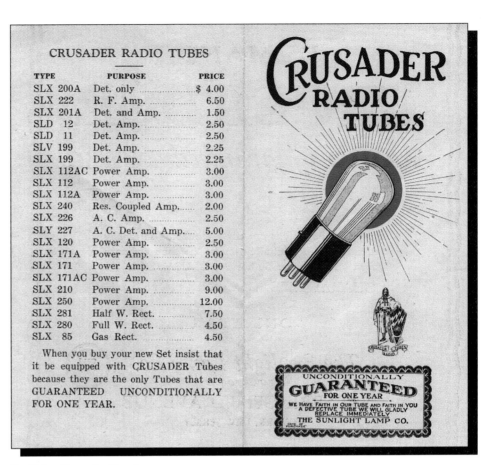

CRUSADER RADIO TUBES		
TYPE	PURPOSE	PRICE
SLX 200A	Det. only	$ 4.00
SLX 222	R. F. Amp.	6.50
SLX 201A	Det. and Amp.	1.50
SLD 12	Det. Amp.	2.50
SLD 11	Det. Amp.	2.50
SLV 199	Det. Amp.	2.25
SLX 199	Det. Amp.	2.25
SLX 112AC	Power Amp.	3.00
SLX 112	Power Amp.	3.00
SLX 112A	Power Amp.	3.00
SLX 240	Res. Coupled Amp.	2.00
SLX 226	A. C. Amp.	2.50
SLY 227	A. C. Det. and Amp.	5.00
SLX 120	Power Amp.	2.50
SLX 171A	Power Amp.	3.00
SLX 171	Power Amp.	3.00
SLX 171AC	Power Amp.	3.00
SLX 210	Power Amp.	9.00
SLX 250	Power Amp.	12.00
SLX 281	Half W. Rect.	7.50
SLX 280	Full W. Rect.	4.50
SLX 85	Gas Rect.	4.50

When you buy your new Set insist that it be equipped with CRUSADER Tubes because they are the only Tubes that are GUARANTEED UNCONDITIONALLY FOR ONE YEAR.

Sunlight Crusader Brochure

Supertron 201-A, 1925
Supertron Manufacturing Co., Inc., Hoboken, N.J. (Joe Knight Collection)

Sylvania S99, 1925
Sylvania Products Co., Emporium, Pa. (Joe Knight Collection)

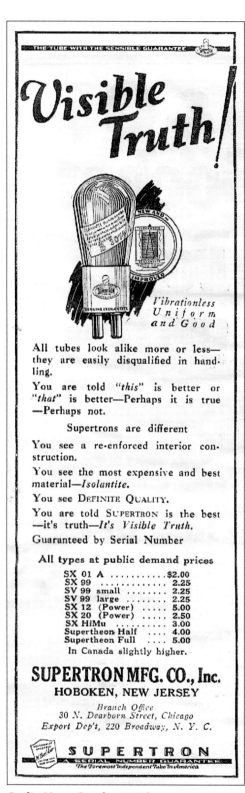

Radio News, **October 1926**

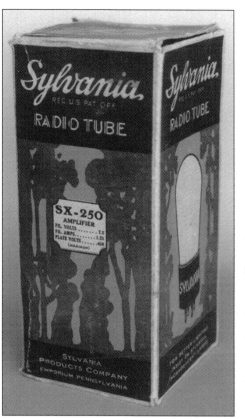

Sylvania SX-250
Sylvania Products Co., Emporium, Pa.
(Kenneth and Debra Krol Collection)

Sylvania 201-A
Sylvania Products Co., Emporium, Pa.

Sylvania 01A
Hygrade Sylvania Corporation,
Emporium, Pa.

Televocal TC. 201A
Televocal Corporation, New York, N.Y.
(Clyde Watson Collection)

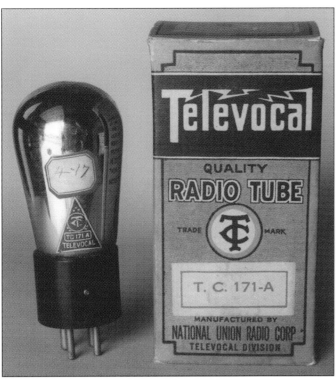

Televocal T.C. 171-A
National Union Radio Corporation, New York, N.Y.

Radio News, **November 1929**

Thoria 201-A
The Thoria Tube Company,
Middletown, Ohio

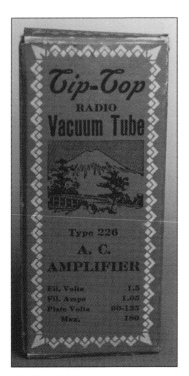

Tip-Top 226

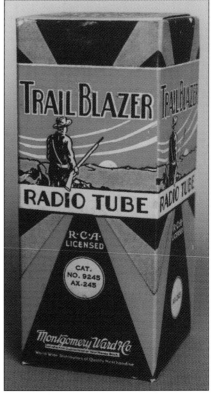

Trail Blazer AX-245
Montgomery Ward & Co., Chicago,
Illinois (Clyde Watson Collection)

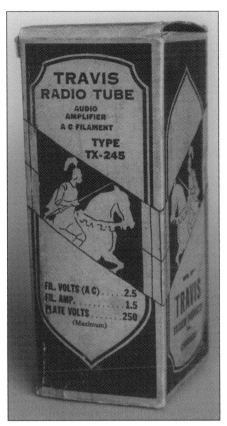

Travis TX-245
Travis Vacuum Products, Inc., Chicago, Illinois

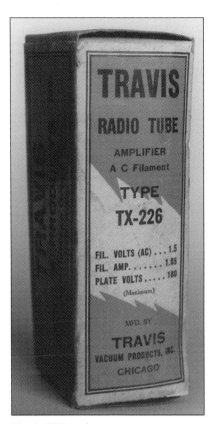

Travis TX-226
Travis Vacuum Products, Inc., Chicago, Illinois

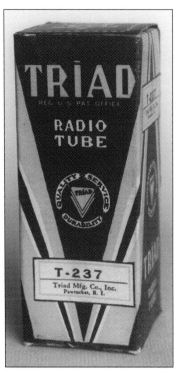

Triad T-237
Triad Manufacturing Co., Inc., Pawtucket, R.I.

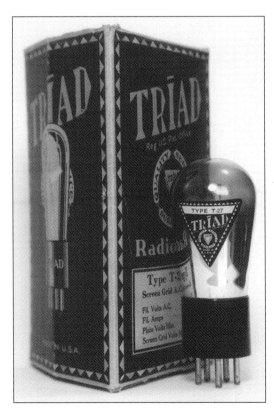

Triad T-27, 1929
Triad Manufacturing Co., Inc. Pawtucket, R.I. The early Triad cartons were triangular shaped. (Joe Knight Collection)

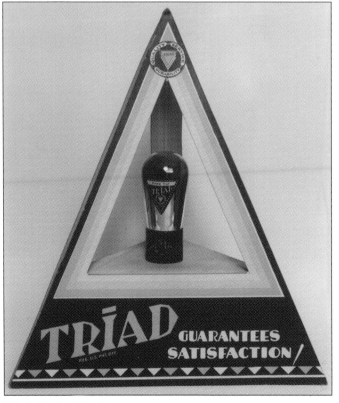

Triad Dealer Display
Triad Manufacturing Co., Pawtucket, R.I. (Joe Knight Collection)

Guaranteed QUALITY...
...*that reduces service calls!*

HOW often the service calls that follow a sale overbalance the money you've made on it! TRĪAD quality stops this dangerous leak in your profits. When you sell a TRĪAD Tube, you're sure of the satisfaction it will give. You're sure also that it will still be giving the same trouble-free performance long months afterward. TRĪAD quality is *insured!* A printed certificate, accompanying each tube, *guarantees* a minimum of six months' perfect service—or a satisfactory adjustment will be made. *Cut your service calls to a minimum—*stock TRĪAD trouble-free tubes! They sell faster and easier, they assure customer satisfaction and dealer protection. They represent your greatest profit opportunity!

TRĪAD MANUFACTURING CO., INC.
PAWTUCKET, R. I.

Call your jobber or write us direct for complete TRĪAD information.

Tune in on the TRĪADORS every Friday evening, 8 to 8:30 P.M. Eastern Standard Time, over WJZ and associated NBC Stations.

TRĪAD
INSURED
RADIO TUBES

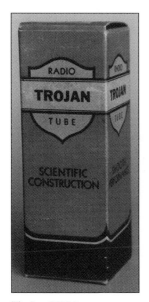

Trojan V199

Troy TX 281

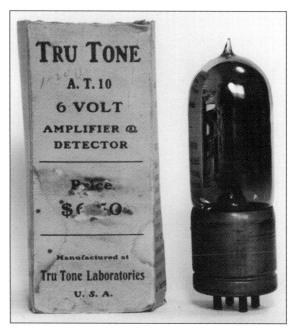

Tru Tone A.T. 10, 1924
Tru Tone Laboratories (Joe Knight Collection)

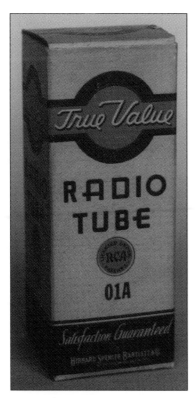

True Value 01A
*Hibbard Spencer Bartlett & Co.,
Chicago, Illinois*

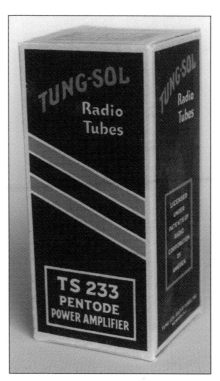

Tung-Sol TS 233
Tung-Sol Radio Tubes, Inc., Newark, N.J.

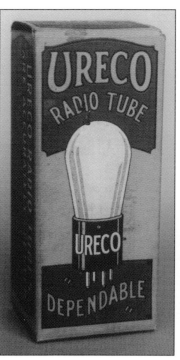

Ureco 201 A, 1926
United Radio & Electric Corporation, Newark, N.J. (Joe Knight Collection)

Ureco X-201A, 1926
United Radio & Electric Corporation, Newark, N.J.

Ureco GoldenTone
United Radio & Electric Corporation, Newark, N.J. (Kenneth and Debra Krol Collection)

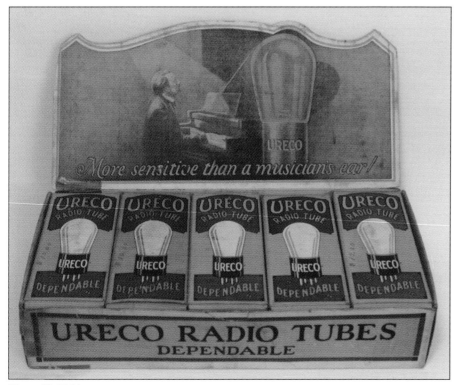

Ureco X-201A
United Radio & Electric Corporation, Newark, N.J. Shown in dealer's counter display. (Joe Knight Collection)

USA Lite UY227
United States Electric Manufacturing Corp., New York, Chicago (Larry Daniel Collection)

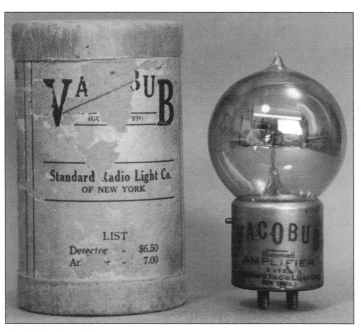

Vacobub Amplifier
Standard Radio Light Co., New York, N.Y. (Joe Knight Collection)

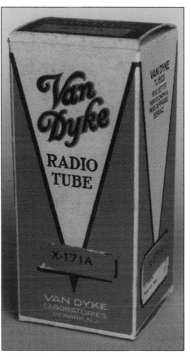

Van Dyke X-171A
Van Dyke Laboratories, Newark, N.J.

Van Horne 5V-A
The Van Horne Co., Franklin, Ohio. Equivalent to UV-201A. (Joe Knight Collection)

Van Horne 199
The Van Horne Co., Franklin, Ohio

Van Horne 227
The Van Horne Co., Franklin, Ohio (Larry Daniel Collection)

You'll Note a Difference

when Van Horne tubes are used throughout your set

Improvement in your reception is entirely up to you—equip your set with Van Horne Tubes throughout —then tune in distant stations with the same ease as you do nearby ones - then listen to the radio programs just as they are being broadcasted.

There will be a new tone quality—a surprising richness and softness of tone that means many hours of radio happiness.

Then too~they'll last clear thru the season

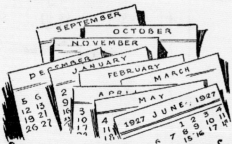

The Van Horne Guarantee

Because of the individual care in the making and testing of each Van Horne Tube a very liberal guarantee is possible.

Should the user find any Van Horne Tube unsatisfactory for any reason it will be immediately replaced. This broad guarantee fully protects the user—insuring the best possible reception.

These Two Unusual Tubes Will Make a World of Improvement in Your Reception

FOR ALL BUT THE LAST AUDIO STAGE
Van Horne Cushion Base Tubes

In every receiving set there is vibration. While barely perceptible it builds up through the various stages of the set resulting in howling, unnatural tone quality and otherwise spoiling reception.

The reason the Cushion Base Tube makes such a wonderful improvement in reception is because the tube is "Cushioned." This means that all vibration is absorbed, giving an unusual softness and fullness of tone of reception that follows the elimination of vibration.

To improve reception order your set from your dealer today.

FOR THE LAST AUDIO SOCKET
Van Horne Adapted Mogul 5 VCX Power Tube

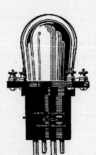

The Mogul 5 VCX is a double capacity power tube—specially to carry all of the signal to the speaker without distortion or loss of signal quality.

Because it can be used in your set without a change in wiring it is an addition that you should make for that increase in volume and improvement in reception that is absent when an ordinary tube of unsufficient capacity is used.

Ask your dealer to demonstrate this remarkable tube to you.

THE VAN HORNE CO., INC.,
1001 CENTER STREET, FRANKLIN, OHIO

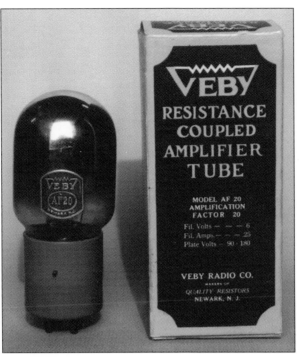

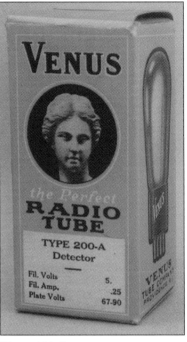

Van Horne 281
*Van Horne Tube Co., Franklin, Ohio
(Larry Daniel Collection)*

Veby AF 20
*Veby Radio Co., Newark, N.J. 6 V. filament, for resistance-
coupled amplifiers.*

Venus 200-A
*Venus Tube Company, Providence,
R.I. (Larry Daniel Collection)*

Via Radio UX 201-A

Vogue AX 227
*Allan Manufacturing Co.,
Arlington, N.J.*

Volutron 201-A

Volutron 201-A
(Clyde Watson Collection)

Vox UX-201 A
Munder Electrical Co., Springfield, Ma.

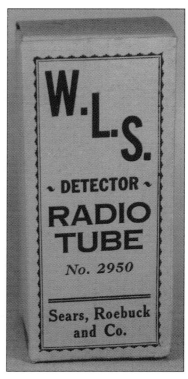

W.L.S. 2950
Sears, Roebuck and Company, Chicago Illinois. Equivalent to UX-200. (Larry Daniel Collection)

WLS 2968
Sears, Roebuck and Company, Chicago, Illinois. Equivalent to UX-171.

WLS 4616
Sears, Roebuck and Co., Chicago, Illinois. Equivalent to UX-171A. (Joe Knight Collection)

W.L.S 2976
Sears, Roebuck and Co., Chicago, Illinois. Equivalent to UX-201A.

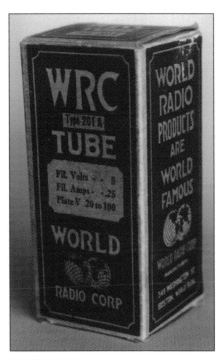

WRC 201 A
World Radio Corporation, Boston, Ma. (Clyde Watson Collection)

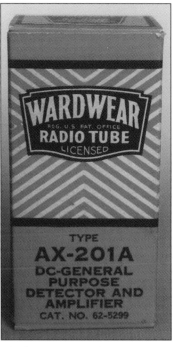

Wardwear AX-201A
Montgomery Ward & Co., Chicago, Illinois

Wearwell UX 201A
The Star Square Auto Supply Co., St. Louis. Missouri (Larry Daniel Collection)

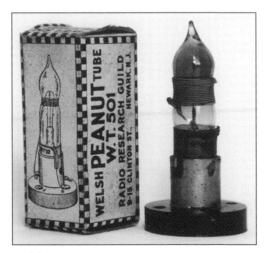

Welsh W.T. 501, 1923
Radio Research Guild, Newark, N.J. The wire wound around the outside of the glass bulb is the grid. (Joe Knight Collection)

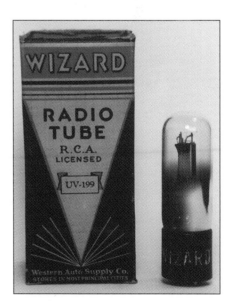

Wizard UV-199
Western Auto Supply Co. (Joe Knight Collection)

Wizard WZX 199
(Joe Knight Collection)

The Wonder B 200
West Factories, Newark, N.J. (Joe Knight Collection)

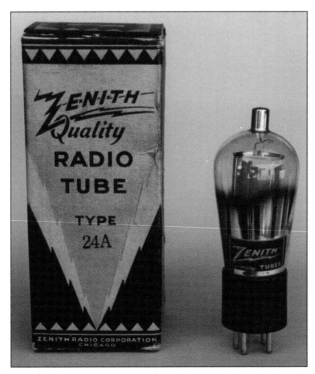

Zenith 24A
Zenith Radio Corporation, Chicago, Illinois (Bill Lettow Collection)

Zetka Z201A
Zetka Laboratories, Inc., Newark, N.J.

201/A BRAND NAMES

The type 201/A vacuum tube was first introduced in 1920 and became the most widely used radio tube throughout the 1920s. Several years ago Bro. Patrick Dowd, W2GK, editor of "The Vacuum Tube" column that appears regularly in the *Old Timer's Bulletin* (Antique Wireless Association) conducted a survey of the readership in an attempt to determine the number of different brand names that these tubes have been sold under. The list has been updated and reprinted in the OTB several times over the years, and last appeared in May of 1995 with an amazing total of 513 brands. Several more brands have since been added, and the current list appears here with Brother Pat's permission.

A

ACE
ART
Acme
Action
Advance
Aerodyne
Airhawk
Air-King
Airline
Airmaster
Airtron
Ajax
Aladdin
Allan - A
Alltron
America
American
Amertron
Amor
Ampladyne IV
   (Germany)
Amplifier
Amplitron
Anylite Electric
A-P
A-P Labs
Alpha
Apex
Apex Audiotron
Apollo
Aratron
Archatron
Arco
Arco-tron
Arcturus
Arion
Aristocrat
Arlington
Armor
Armur
Arotron
Atlantic
Atlantic-Pacific
Atlantis
Atlas
Atron
Audiotron
A.W.A. (Australia)
A-1 Wonder

B

Barium
Beacon
Belco

Belknap
Belltone
Bestone
Betta-Tone
B-M
Blazing Train
Blo-Pruf
Bluebell
Blue Bird
Blue Grass
Blue Ribbon
Blue Streak
Bluetron
Boehm
Bond
Boone
Bosch Radio
Bostron
Brede
Bremco
Bright Star
Brilliantone
Buck
Bull Dog
Bumblebee
By Heck

C

Capex
Cascade
Cecilian
CeCo
Champion
Chesterfield
Chibar
Clarion
Claritone
Claritron
Clearatone
Cleartone
Cleartron
Clover
Coast To Coast
Columbia
Comet
Commander
Concert
Concert Master
Continental
Conqueror
Cornell
Coronado
CPC
CR
Crescent
Crescer

Crosley
Croydon
Crusader
Crystal
Cunningham
Curtiss Energy Co.
Cymotron (Japan)

D

Daven
Davis
Defender
DeForest
Delco
Denmark
Detectron
Diamond
Diana
Diatron
Dilco
Dilling
Dixie
Dis-X-Ron
Dollar
Don (Japan)
Donle-Bristle
Double Life
DRC
Duotron
Duovac
Duraco
Duratron
Duro
Durotron
Dutch Radio Valve
Duval
Dynamic
Dynatron
Dynetron

E

Eagle
Echotron
Eclipse
Econotron
Econotube
EDLO
Electrad Triode
Electron 6000
Elektron
Elitetron
EMCO
Emperor
Empire
Empire-Tron

Endurance
ERC
ERLA
Esetron
Essex
Eveready-Raytheon
Everedy
Everest World-Top
Excel
Excello-Tron

F

Falck
Falco
Farad
Fearnola
Federal
Federal Supreme
Field
Field's
Flair
Franklin
Fultone

G

Gebhart
Gem
General Electric
General Electric
   (Canada)
Gibraltar
Globe
Gloria
Gold Bond
Goldentone
Gold Seal
Goldtone
Good Luck
Goode Tube
Grand Tone
Grant's Special
Green Seal
Greentron
Greylock

H

Halcyon
Harmonique
Harp
Harvard
"H" Brand Blue
Heliodyne (Germany)
Henderson
Hercules
H.H. Co.

Hi-Constron
Highest Grade
High-Mu
Hi-Mu
HO
HOY
HR-O
Hygrade
Hylon
Hytron
Hy Vac

I

Ideal
Imperial
Imperialtron
Iroquois

J

Jaeger
JA REG
J&R
Jayes
Jewel
Jim Brown
Jove
JRC

K

Kazoo
Keen Tone
Kelly
Kelvin
KenRad
Keystone
Kilbourne & Clark
Kinodyne
Kleen Tone
Kleer-Tone
KMA
KR
Knight
Kwik-lite

L

Lafayette
LaSalle
Leader (The)
Leco
Lectron
Lee
Legion
Leradion
Lewis De Luxe

Lestron
Liberty
Lincoln
Little Giant
Live-Tone
Livestone
Lord Baltimore
Loudspeaker
Luro
L.W. Cleveland Co.
Lyric

M

Macleco
Macy's
Magestic
Magictron
Magnadyne
Magnatron
Magnavox
Magnet
Marathon
Mark's
Marle
Marvin
Marwol
Master
Mastertone
Maston
Matchless
Maximus
Maximus Quality
MECO
Mello-Tron
Melophonic
Milo
Milo Blue Bell
Miraco
Mitchell
Mizpah
Monotron
Mozart
Murdon
Music Master
Musiktron
Musselman
Muzada
Myers (01A)

N

Nathans
National
National Products
National Union
Navigator
Neonlite
Neontron
Neptron
Neptune
Neutrone
Northome
Novatron
Novo
Novo-Tron
Nultone
Nu-Tro
Nutron

O

OK
OKR
Operatone
ORC Detron

Oreco
Oriole
O&T Electric Co.
O-T Olney Co.
Oscillector
Ozarka

P

P.V.F.
Pace Setter
Panama
PAR
Packard
Paramount
Peak
Pearson
Peerless Radio Valve,
    Ltd. (Canada)
PeerTron
Penn Wave
Pep Boys
PepTone
Perfect
Perfectron
Perfectone
Permatron
Perryman
Phenix
Philco
Philips (Holland)
Philotron
Philmore
Phonotron
Pilot
Pilotron
Pingree
Pioneer
Pla-Mor
Plarite
Playto
Platron
Playtron
Pliotron
Powertone
Powertron
Premier
Prextro
PRL
Princeton
Pur-A-Tone
Puratron

Q

QRS
Quadrotron
Quaker
Qualitron
Quality
Q.V. Tron

R

RA-DEX (Germany)
RE-DEX
RA-DEXTRON
Radio Knight
Radios
Radiotone
Radiotron (GE, RCA,
    & Westinghouse)
Radiotune
Rainbow
Ranger
Rayo

Ray-O-Vac
Raytheon
Raytron
Rayvac
RCR
Real-Tone
Reception
Receptron
Red Printing
Regal
Reliance
Remarc
Research
Rextron
Ritetone
Robun
Roice
Rolls Royce
Rowtron
Roxy
Roy
Royal
Royal Blue
Royale
Royalfone
Royaltron
Roytron
RSK
RVC (Canada)
RY-TONE

S

Samson
Savoy
Schickerling
Sea Gull
Shamrock
Shelby
Shepherd
Signal
Silver
Silver Domino
Silver King
Silver Shield
Silvertone
Silvertron
Simplex
Skyline
Sky Sweeper
Slagle
Solartron
Songbird
Sonora
Sonatron
Sovereign
Sparton
Speed
Standard
Standard Black-Base
Stanton
Star (NJ)
Star-Ko
Star Tube (Holland)
Sterling
Stewart-Warner
Stratford
Strongson
Sturdy
Sunlight
Sunlight Crusader
Super Air Castle
Super Airline
Supercraft
Super-Het

Superior
Superior Tone
Supertone
Supertron
Supertron Precision
Surety
Sylfan
Sylvania

T

Tan Box
Taylor V.T. Co.
Tayrad
Tectron
Teletron
Televocal
Teltron
The Perfect Tube
Thermionic
Thermatron
Thoria
Thorolla
Thurman Thermionic
Tipless-Rytone
Titan
Titania
Tobe
Trail Blazer
Trav-ler
Traveler
Trego
Triad
Triple Tone
Trojan
Troy
Tru Tone
True Blue (Brightson)
True Value
Tung-Sol
T-X

U

Ultra
Unitone
Unitone No Bee
Universal
Ureco
USALITE

V

Vacobub
Vactron
Vancy
Van Dyke
Van Horne
Veby
Venus
Veribest
Vesta
Via Radio
Viking
Vim
Vista
Vitavox
Vogue
Vogue Non-Pareil
Voltron
Volutron
Vox
Vultron

W

Walco
Wards
Wardwear
Wasco
Wavetone
WDRC
Wearwell
Weile
Weltone
WCA (Wireless
    Corp. of America)
WES
Western Electric
Western Lab
Western Quality Tube
Westinghouse
Westinghouse (Canada)
Weston
Westron
Williams Certified
Wizard
WLS
Wonder
Wonder-Tone
World
WRC

Y

Yankee

Z

"Z"
Zenith
Zetka

SELECTED BIBLIOGRAPHY

Fathauer, George H. *Electron Tube Resume*. Antique Electronic Supply. Tempe, Arizona, 1993

Magers, Bernard. *75 Years of Western Electric Tube Manufacturing*. Antique Electronic Supply. Tempe, Arizona, 1992

Sibley, Ludwell. *Tube Lore*. Ludwell A. Sibley, 1996

Stokes, John W. *70 Years of Radio Tubes and Valves*. 2nd Ed. Sonoran Publishing. Chandler, Arizona, 1997

Tyne, Gerald F.J. *Saga of the Vacuum Tube*. Antique Electronic Supply. Tempe, Arizona 1977